BE YOUR OWN
CHIMNEY SWEEP

BE YOUR OWN CHIMNEY SWEEP

by
Christopher Curtis
and **Donald Post**

Photographs by Glen Moody and Dan Lax, courtesy of Black Magic Chimney Sweeps

Line drawings by Cathy Baker

Second printing, September 1979

Library of Congress Cataloging in Publication Data

Curtis, Christopher, 1951–
Be your own chimney sweep.

Includes index.
1. Chimneys—Cleaning. 2. Stoves—Cleaning.
3. Fireplaces—Cleaning. I. Post, Donald, 1948–
joint author. II. Title.
TH2285.C87 648'.5 79-12608
ISBN O-88266-157-4

Contents

Introduction

Many people view chimney sweeping as an age-old art, no longer needed in the hustle and bustle of today's world. Actually, both notions are false. With the return of wood as a popular heating fuel has come the need for periodic cleaning of chimneys. And while it is true that chimney sweeps have been pursuing their calling for a few hundred years, from a broader perspective, sweeps have emerged only recently in the history of domestic fire.

As is true with most of American culture, chimney sweeping has its roots in Europe. It is unclear where it was popularized first, but most legend springs from Great Britain, where the famous climbing boys were "employed" by master sweeps. Little was written about the task of chimney cleaning until the seventeenth and eighteenth centuries, when the hardships of the climbing boys came into public view.

The first apparent use of the term *chimney sweeper* was in the seventeenth century. John Cottington, born in Cheapside, England in 1611, was apprenticed to a Master at the age of eight. By the time he was thirteen, he considered himself "as fully informed in the art and mystery of chimney sweeping as the Master," and he ran away. His invention of the term *chimney sweeper* earned him a place in chimney sweeping history, but it didn't save him from the gallows, where he died in 1656 after a romantic career as a highwayman known as "Mulled Sack."

Climbing boys were used occasionally in such cities as Paris, Edinburgh, Glasgow, Dublin, New York and Philadelphia, but London was the center of the chimney sweeping world. Elsewhere, a rope with a bush or sort of broom tied to the middle was dragged up and down flues to clean them. London, however, had more than its share of zig-zag flues, each flue serving a fireplace in a different room. To clean these flues it was necessary, the English thought, to send a small child into the flue, the average size of which was about nine by fourteen inches.

Small children well fit for this purpose were often purchased under the guise of apprenticeship by the master sweep who himself was also a victim of the times. Typically, a poor father would consent to apprenticing his young son to a master sweep in order that he might learn a trade. Unfortunately, few climbing boys survived their apprenticeships, often succumbing to consumption (tuberculosis) or chimney sweeps' cancer (cancer of the scrotum). The fame of climbing boys comes from the era of child labor throughout England.

It was commonly recognized that the chimney sweeps' lot was not a good one, and as early as 1796 measures were taken to ease their plight. The House of Lords in Parliament was dreadfully opposed to changing the way chimneys were swept, and over the years it stifled many legislative attempts to improve working conditions. This opposition though, did not deter others from trying.

The Society for the Encouragement of Arts, Manufactures, and Commerce offered its gold medal to anyone who "...shall invent and produce to the Society the most simple, cheap, and proper apparatus, superior to any hitherto known or in use for cleansing chimneys from soot, and obviating the necessity of children being employed within the manner now practiced." This offer stood for a number of years and was finally dropped in 1802 only to be revived the following year by a new organization, The Society for Superseding the Necessity of Climbing Boys, By Encouraging a New Method of Sweeping Chimnies, and for Improving the Condition of Children and Others, Employed by Chimney Sweepers.

The prize was first won in 1805 by George Smart for his "machine"—a brush of sorts on extendable handles. An off-shoot of Smart's apparatus is still in common use in the United States. The basine brushes have been replaced with steel, and the bamboo poles are now fiberglass, but the basic idea remains unchanged.

Eventually, machines replaced climbing boys, both out of convenience and pressure from Parliament, but not before the climbing boys had left their mark in medical history. Henry Earle's work, "On Chimney Sweepers' Cancer," published in *Medico-Chirurgical Transactions* in 1823, became the first documentation of occupationally linked disease.

The chimney sweeps of today, I am glad to relate, are a happy lot. The tools they use now are not very different from the ones used in the

1800's, except for the modern dust collector, a special large-volume vacuum essential to today's sweeps. Until very recently having one's chimney swept meant coating one's house with a layer of soot. Today's professional sweep and homeowner alike should have little trouble keeping the mess to an absolute minimum.

The purpose of this book is to advise the homeowner on how to clean a woodstove, fireplace, solid-fuel furnace or boiler, stovepipe, and chimney. One would, of course, have some combination of the above in one's home, and in the interest of brevity, discussion of each entire system has been avoided in favor of part of each system. For example, we have avoided separately discussing a chimney for a woodstove and a chimney for a furnace.

A note about terms. Within this book a *fireplace* will mean an open-front burning appliance, either masonry or prefabricated. A *stove* is a closed burning device, either for heating or for cooking or both. *Stovepipe*, also called *chimney connector* or *smoke pipe*, is the thin-wall pipe connecting the stove to the chimney.

The *chimney* is the part of the system that carries the smoke up to the roof. It may be made of brick, concrete, insulated stainless steel or uninsulated triple-walled pipe. The chimney may also be called the *stack*. Occasionally the term *stack* will include both the chimney and the stovepipe. See the Glossary for other terms used throughout the book.

Cleaning chimneys traditionally has been thought of as a despicable task relegated to those unfortunate individuals doomed to the role of chimney sweep. However, almost anyone who has a chimney can clean it, using a little elbow grease and a lot of soap. Slightly soiled hands are, after all, a small price to pay in return for the millions of Btu's your stove has provided you or for the countless times you've gazed into the flames of your fireplace.

You may find that you don't wish to soil your hands ever again at this task. Or you may find that the despicable task and the unfortunate individual aren't so despicable or unfortunate after all. You may, just for a moment, view the world through the eye of a sweep, and you'll know why we sing. So take heart, and when your hands are black—smile.

Why Chimneys Need Cleaning

Many people wonder why a chimney ever needs cleaning at all, so before we cover how to do it, let's make sure it needs doing. When wood burns, its combustion is never quite complete. Even under very controlled combustion conditions, one is left with ash, but in practical conditions, much more than ash escapes without being burned. Many flammable gases escape as smoke, but more important, a material called *creosote* is given off when wood burns. Creosote is flammable, and if it can be made to burn in the fire, it adds valuable calories to the heat output. However, the ignition temperature of creosote is high, and often much of this creosote escapes up the chimney, carried in smoke in the form of tiny droplets of liquid. Unfortunately, everything that enters the chimney at the bottom does not exit at the top. Part of what is left behind is creosote, which condenses onto the relatively cool surface of the chimney interior.

CREOSOTE FORMATION

Creosote can occur in several forms, ranging from a liquid (which is mostly water) to a crusty, brittle ash. The watery kind is not particularly hazardous, but it does make a mess, especially if it regularly drips from stovepipe joints.

Another form of creosote is a tacky, thick layer not unlike partially dried paint. Sometimes creosote will dry to a glassy-surfaced, exceptionally tough coating. Modern airtight stoves sometimes form this type of creosote.

The most common type of creosote is a dry flaky coating. This kind of deposit is generally formed by a fireplace fire, or by a stove that is not very airtight and has high stack temperatures.

Because creosote *is* flammable, it is wise to keep the amount within your chimney to a minimum. Even the watery kind of creosote can be dried and heated to its burning point. If creosote in a chimney does ignite, the result is a chimney fire. This is what you are trying to avoid. Chimney fires can be brief and harmless or, if they are long or intense, they can burn down the building. Often they render an entire chimney unsafe without producing any visible damage on the chimney exterior. By keeping your chimney clean and firing your stove in a manner to keep creosote formation to a minimum, you can reduce the threat of chimney fires.

How a Log Burns

The process of wood burning seems straightforward, but there are many steps in the conversion of wood to ash. Let's assume that we begin with an air-dried (20 percent moisture content) log of maple, a typical hardwood. We add it to a burning open fire. Immediately the surface begins to absorb heat. The surface temperature rises. As it does, heat is conducted farther into the wood, pyrolizing the wood into gases that escape and burn at the surface. The temperature rises steadily, but the chief change is evaporation of water until the wood temperature reaches 100°C. (212°F.). At about this temperature, the wood continues to absorb heat, but the temperature does not rise until all the moisture has been converted to steam. (This process consumes considerable heat—about 1150 Btu per pound of water, underscoring the desirability of reasonably dry wood).

When all the water has changed phase, the temperature continues to rise, driving out still more water vapor, carbon monoxide, carbon dioxide, formic acid, acetic acid and other compounds. This process is called *destructive distillation*. With the exception of water vapor, these chemicals are created by the heating process; they are not merely "dried" out of the wood. These changes are still *endothermic*—that is, they require the addition of heat, even though many of the gases generated are flammable.

At somewhat over 270°C. (520°F.) many more gases are generated and the reactions become *exothermic*. That is, wood gives off small amounts of heat at this point, even without the presence of fire. Water

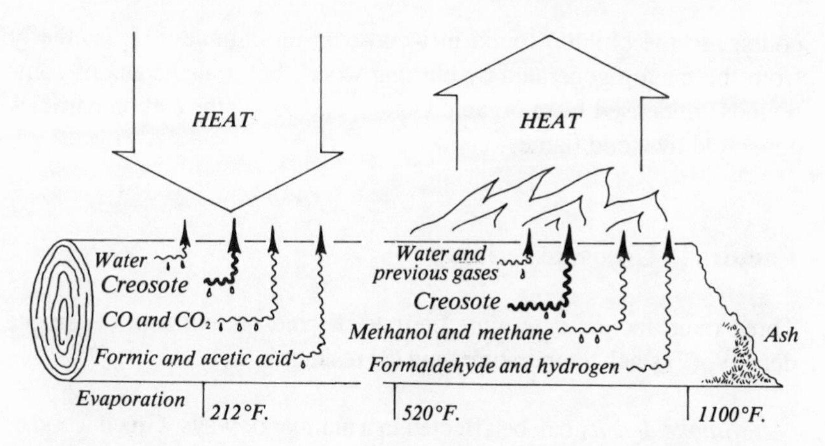

How a log burns.

vapor and the first group of gases continue to be given off, joined by methane, methanol, formaldehyde and hydrogen. Some of these gases are burned in the fire; some escape up the chimney.

At the same time, *wood tar* is being generated. Tar contains hundreds of different compounds—some flammable, some not. It is this tar combined with soot that forms most of the residue coating the inside of the chimney, commonly known as *creosote*. Tar from a wood fire generally does not exist as a gas, but leaves the wood in what is known as *tar fog*—tiny droplets of liquid tar suspended in smoke. Some of this tar fog is burned as it passes the burning region at the surface of the wood, but much of it escapes up the chimney. At about 500°C. (930°F.) the only remaining substance is charcoal. The charcoal continues to burn, providing most of the heat of the fire. Finally, the only remaining thing is ash. Everything else has either been burned or carried away in smoke.

Our assumptions for the process above provided for a log on an open fire, but in an enclosed chamber, such as a stove, particularly an airtight model, with the draft closed—burning is rarely so complete. The combination of a slow, smoldering flameless fire and low stack temperatures and smoke velocities creates higher rates of creosote formation.

What is Creosote?

Strictly speaking, creosote is a specific chemical compound ($C_8H_{10}O_2$), but the word also refers to a wood preservative derived from coal and, of course, to the buildup found in wood-burning chimneys. It is chiefly from the tar fog generated by burning wood, but it also contains compounds condensed from organic vapors, and soot—the carbon particles generated by wood flame.

Factors in Creosote Buildup

Three main factors determine the rate of creosote buildup: (1) smoke density, (2) stack temperature, and (3) residence time.

Smoke density can be affected in a number of ways. Green wood or softwoods commonly yield high amounts of smoke, particularly if they are burned where there is not enough oxygen. This is often the case in a closed air-tight stove. There also are commonly large amounts of smoke when fresh wood is added to hot coals. Smoke can be made less dense by adding more air to the combustion process or by adding *secondary air* or *dilution air* to the stack.

Secondary air means that warm air from the house is being fed into the stove and being lost up the chimney. It is unclear that adding secondary

A Riteway *wood furnace. Cutaway version shows path of primary air (bottom), which is controlled by thermostat. Secondary air enters to help burn combustion gases.*

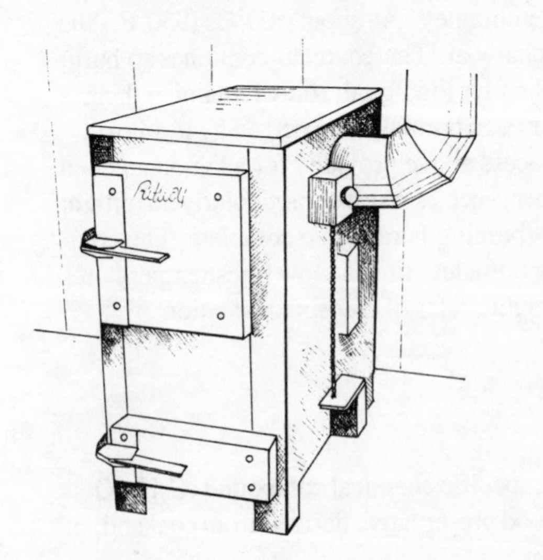

air has any net positive effect because of its cooling effect on the stack. The National Fire Prevention Association states that dilution air actually increases the rate of creosote deposition. Unless your stove has a sophisticated secondary air inlet, I would steer clear of either secondary air or dilution air. While secondary air may be good in theory, it is unlikely that unburned gases with high ignition temperatures will be ignited even if oxygen is present. It is obvious that the amount of *tar* in the smoke will have a direct bearing on how much winds up on the chimney lining.

Stack temperature is another very important factor in creosote buildup. A cold chimney will cool vapors enough to cause them to condense on the chimney lining surface, depositing creosote. The condensation often includes water, which combines with the tar to form a thin, watery sort of creosote, often called *pyroligneous acid*. This is the stuff you've probably seen running down the outside of some stovepipes. Creosote without water usually is too viscous to flow easily.

Keeping stack temperatures high will slow or even eliminate creosote formation. Unfortunately, doing this means sending large amounts of heat from the fire up the chimney. Heating engineers call this *stack loss*. There is a direct tradeoff between stack loss and creosote buildup. A stack temperature of 300° C. (572° F.) will eliminate nearly all creosote, but it means very high stack loss. A stack temperature of 50° C. (122° F.), by contrast, would mean very low stack loss but very fast creosote deposition.

A stack temperature of at least 100° C. (212° F.) probably is desirable for three reasons. First, at that temperature water will remain a vapor. It will exit the chimney top rather than condense on the chimney liner and combine with other deposits to run down the stovepipe and drip on your carpet.

Second, 100° C. (212° F.) is a good compromise between creosote and stack loss. You'll still get some of both, but not too much of either. Incidentally, an easy way to test for the 100° C. mark is to sprinkle a few drops of water on the stovepipe. If it hisses, you've got at least 100° C.

Third, a temperature of 100° C. is usually needed to maintain a good draft. This draft is necessary for any stove or fireplace to keep it burning and to eliminate smoking into the house. (Stack temperatures of less than 100° C., however, will often be high enough to maintain good draft.)

A convenient surface thermometer is available called *Chimgard*™ for stovepipe and stove that will show operating temperature. Buy it at stove stores or order by mail ($10) from Black Magic Chimney Sweeps, Stowe, Vermont 05672.

Watery creosote running down an outside uninsulated chimney. Using an insulated stack avoids this problem and keeps creosote to a minimum.

Stack temperatures are affected by a number of things. Mode of burning, type of fuel and its moisture content, secondary or dilution air allowed in, damper settings, type of chimney, temperature surrounding the chimney, and the use of heat reclaimers all affect the stack temperature. Some of these conditions can be controlled by the stove operator, so creosote buildup can be slowed by careful control of the stove.

More important considerations may be stack construction and ambient (surrounding) temperatures. Creosote forms in a masonry chimney on the outside of the building fast because it is a very good *heat sink*. That is, the chimney will absorb a lot of heat from the cooling flue gases. By contrast, an insulated chimney, such as *Metalbestos* brand, located mostly within the heated portion of the house, will keep the flue gases warm and form relatively little creosote. (Any chimney being fired from a cool state must be warmed, however, and so it must at least pass through a temperature range that encourages creosote formation.)

Long lengths of stovepipe can add considerable heat to the house, but they also cool the flue gases, increasing creosote buildup and water condensation. Heat reclaimers designed to be inserted in stovepipe do reclaim more heat from the exiting smoke, but they also lower stack temperatures, accelerating creosote formation. Unfortunately, by keeping creosote buildup to a minimum with high stack temperatures, we also experience high stack loss. You can't have your cake and eat it too.

Residence time, the third major factor contributing to creosote buildup, is the amount of time the smoke "resides" within the system—that is, within the stove, stovepipe, and chimney. A short residence time means that the smoke moves quickly out of the system, aided by a good air supply. This also means that the smoke will be diluted. A short residence time means high stack temperatures—and heat losses—since there is little opportunity for heat to be absorbed into the home. A long residence time allows ample opportunity for heat to be transferred out of the system, lowering the stack temperature and increasing creosote buildup.

Residence time depends on chimney length and diameter, stack height, and flue gas velocity. *Chimney length* generally is fixed or can be affected only slightly. *Flue gas velocity,* however, can be affected by chimney draft and by damper settings. As with stack temperatures and smoke density, the damper setting still is the main way to control residence time and the resulting creosote buildup.

Another minor factor—*turbulence*—affects creosote buildup. Turbulence introduced into flue gases by elbows in stove pipe or by dampers is said to increase the rate of buildup. This effect is very small when considered beside the major factors.

Slow Burning Means More Creosote

Any wood fire produces creosote. There is an almost direct trade off between the amount of heat recovered and the rate of creosote accumulation. If your system is very energy efficient, and recovers most of the heat of combustion, you can count on fast rates of buildup. If you are willing to sacrifice some efficiency, you can reduce buildup and clean your system less often.

CHIMNEY FIRES

Most people who burn wood have at least heard of chimney fires and the threat of chimney fires is the best reason to clean chimneys. There is always the direct danger that the creosote that has been deposited will ignite and burn, heating the chimney to dangerous levels. Let's examine how this happens.

Creosote has been deposited in the chimney by the process we just described. Probably it was deposited as a liquid in combination with some water. The water has evaporated, leaving a shiny hard coating not unlike black varnish (and no easier to remove). As the chimney warms up, either from subsequent fires or from hotter combustion temperatures, the creosote may be altered substantially, both physically and chemically. Volatile compounds evaporate. When heated further the creosote will *pyrolize* to form yet another generation of compounds. In this process, the physical form of creosote changes to a bubbly surface and then to a flaky, dull-finish crusty material.

As the temperature continues to rise, the creosote finally ignites. It increases the draft of the chimney, feeding itself more oxygen and burning progressively hotter. Temperatures high enough to ignite creosote may occur only at the very bottom of a chimney, but when the creosote at the base has been ignited, it is easy for fire to spread up the entire chimney. As more and more of the creosote burns, the suction increases. More oxygen enters, causing an incredibly hot fire.

Temperatures in chimney fires are not well documented, and depend on several factors. Temperatures of 1000° C. (1832° F.) to 1500° C. (2732° F.) are not uncommon, and I have heard figures of as high as 2500° C. (4532° F.). Chimney fires of this intensity, even if brief, can heat masonry chimneys hot enough to ignite adjacent combustibles.

Creosote may be deposited in a shiny, hard form, left. Subsequent reheatings may change it to a thick flaky crust.

Even small cracks in masonry can allow fire through to wood or other combustibles.

Chimney Cracks

Masonry chimneys are very heavy by nature—a 20-ton chimney is not uncommon—and some settling after construction is inevitable. As a result of this settling, cracks often will develop. The point at which a chimney exits the roof is also likely to develop cracks. Wind often sways chimneys and if insufficient space for this motion is left in the roof, cracks may result. Any of these could allow fire to escape.

Stainless steel insulated chimneys are quite resistant to deterioration by fire, but in hot chimney fires, they too can warp and develop cracks at the joints.

In stovepipe, chimney fires are especially dangerous because the thin walls of the pipe heat up much faster than bricks or masonry. In addition, they can burn through as well as reradiate great amounts of heat to nearby combustibles. It is not uncommon for stovepipe to be heated red hot by a chimney fire. If it changes color to even dark red, its temperature is at least 475° C. (885° F.). If it becomes cherry red, its temperature is about 750° C. (1380° F.), easily hot enough to burn wood near the pipe. See Table 1.

Table 1. The Approximate Color of Glowing Hot, Solid Objects.

Appearance	*Temperature* °F	°C
No emission detectable	Less than 885	Less than 475
Dark red	885–1200	475–650
Dark red to cherry red	1200–1380	650–750
Cherry red to bright cherry red	1380–1500	750–815
Bright cherry red to orange	1500–1650	815–900
Orange to yellow	1650–2000	900–1090
Yellow to light yellow	2000–2400	1090–1315
Light yellow to white	2400–2800	1315–1540
Brighter white	higher than 2800	higher than 1540

Adapted from D. Rhodes, *Kilns* (Philadelphia: Chilton Book Co. 1968), p. 170.

Creosote Corrosion

Creosote is acidic and deteriorates stovepipe. A flue fire or chimney fire may be the proverbial straw that breaks the camel's back and spews fire into the house. Creosote also corrodes steel and mortar. At the high temperatures encountered in a chimney fire, its corrosiveness increases dramatically. Ceramic tiles used to line masonry chimneys and stainless steel are very resistant to corrosion, but at high temperatures, they too corrode. The heat expands and cracks masonry. If a chimney is improperly constructed, there may not be room for expansion of the lining, causing it to crack or break up.

Broken Tiles

If unwitting persons attempt to extinguish the fire by pouring water into the chimney, the rapid cooling and contraction can break tiles. Properly constructed flue tiles have no mortar between the joints; rather the joints are ''belled'' with mortar around the joints (see the illustration). How-

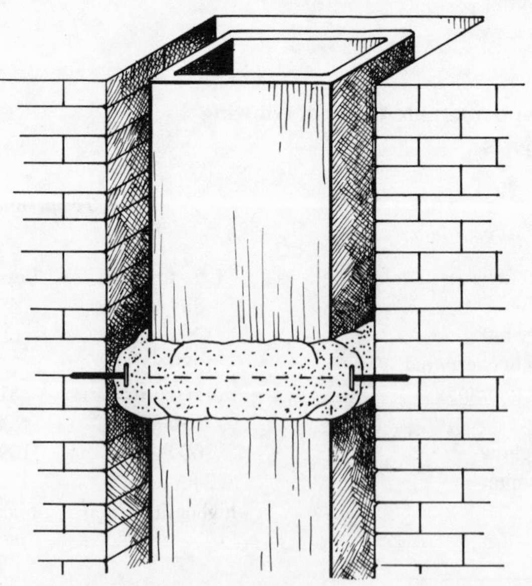

Flue tiles are mortared on the outside *of the joint.*

ever, not all masons follow this guideline. Mortar exposed to chimney fire temperatures can break down and leave gaps between flue tiles. In the case of unlined chimneys, the deterioration experienced in a chimney fire is even more pronounced. Bricks should be able to stand chimney fire temperatures, but they too expand, encouraging cracks within the brick and at the brick and mortar interface. Then much more mortar is exposed to the direct heat of the fire.

Vibration and Zinc Vapor

Occasionally, a flue fire will cause violent shaking of the stovepipe, making stovepipe that is not securely fastened a real hazard. What happens in this case is that a chimney fire begins but is oxygen-starved. It extinguishes itself for lack of oxygen, but continues to draw air into the hot pipe. When enough oxygen is present to allow burning to continue, the fire bursts forth, again only to extinguish itself. This burn/smolder cycle can be repeated several times each second, setting up vibration in the pipe and stack. Unfastened or particularly corroded stovepipe become real hazards in this instance, as the stovepipe may shake itself loose, showering the bewildered inhabitants with red hot steel and flaming creosote.

If galvanized stovepipe has been used, the galvanization (which is mostly zinc) can vaporize. Zinc vapor on the inside of the flue is of little danger, as the vapor is carried up the chimney, but galvanization on the *outside* of the pipe can release zinc vapor into the living area.

WHAT INSURANCE COMPANIES SAY

Every year thousands of homes are destroyed or seriously damaged by chimney fires. Understandably, insurance companies are concerned. Some companies have cancelled insurance policies upon learning that a wood stove has been added to a residence.

To protect your policy and your home, make sure that your wood-burning system is safe—and inform your insurance company. Invite them to inspect your installation. Their advice and approval will protect you and your investment.

WHAT TO DO IF YOU HAVE A CHIMNEY FIRE

1. Call the fire department. You may not need them, but you are better safe than sorry; by the time it becomes apparent that you do need them, it will be too late.

2. Try to slow or stop the fire by depriving it of oxygen. In an airtight stove this is relatively simple: close all draft controls. In a stove that is not airtight, this won't put the fire out, but it may cut the intensity.

In a fireplace, you can cover the opening with a piece of plywood or similar material, but beware of strong suction created by the chimney fire. A wet blanket can be draped over your fireplace screen to seal the opening, but it must not be used without something to keep it from being sucked into the fire. *Do not close the damper,* since the smoke from the fire in the fireplace will come into the room.

3. Use a fire extinguisher. The best thing to use is a flare-like fire extinguisher designed specifically for chimney fires. They are available from fire equipment companies and many chimney sweeps. *Chimfex* is one brand name. These extinguishers are great to have on hand just in case and cost under $10.00. To use, break off the top and strike the end of the stick with the broken off top, just like a road flare. Then place it inside the stove or on the smoke shelf of the fireplace.

Regular CO_2 fire extinguishers can also be used. Simply direct the nozzle into the chimney. A few applications may be necessary.

Do not pour or spray water into the chimney. The rapid contraction caused by cooling will almost certainly break tiles.

4. Check the chimney temperature. Feel the surface of the chimney anywhere you can put your hand on it. If it gets too hot to touch, summon help immediately. Continue checking repeatedly—a chimney fire may last almost an hour, although most are over in minutes.

5. Check the outside. Sparks and embers will be blown out the chimney top and can ignite roofs or nearby brush. Go outside and see where they're landing.

A dangerous installation. The elbows of thin-walled stovepipe build up creosote quickly, and a chimney fire would rapidly ignite the newspapers and unshielded flaking wallpaper a few inches away.

6. When it's over, sweep the chimney and inspect for damage. Many people think that a chimney fire cleans the chimney. It doesn't. In fact, the ash left from a chimney fire may have expanded to take up more of the flue area than the creosote did. It also often falls to base of the chimney, forming a major block. When the chimney has been cleaned, carefully inspect it for damage. Has the fire cracked any tiles, or knocked out mortar?

7. Resolve to clean your chimney more often. Chimney fires cannot happen unless there is enough creosote to fuel them. By cleaning more often you remove the fuel and eliminate the cause of the fire.

The Second Fire

It is often said that the second chimney fire is the one to beware of. The implication is, of course, that the chimney was damaged by the first fire and the second one will be much more dangerous to the house and occupants. Burning debris shooting out the top of the chimney can also present a hazard, particularly if the roof is wood or if the surroundings are dry and flammable.

OTHER PROBLEMS CREOSOTE CAUSES

Creosote is undesirable, not only because it is fuel for chimney fires, but for several other reasons. It decreases the effective flue diameter of the stack. This reduction is most dramatic in smaller stacks. For example, a six-inch pipe with a one-half inch buildup of creosote loses 30 percent of its area. The result is lowered draft, with accompanying slower burning in the stove. The stove may smoke when starting or when the fuel door is opened to add fuel, or in windy conditions. The initial buildup inside the pipe also speeds up the rate of creosote deposit thereafter.

Creosote also is an excellent insulator. Buildups of as little as two millimeters can cause heat transfer efficiency drops of between 10 and 15 percent. This is true inside a stove as well as in a stovepipe. Stovepipe often can contribute as much as 20 percent of the recovered heat; thus keeping it clean can add considerable heat to the house. In woodburning furnaces and boilers, heat transfer efficiency is very important, so keeping them clean is essential.

Practically speaking, the thing that is often the most annoying about creosote deposits is the mess they make in the home. This is particularly true if the creosote is watery, since creosote will generally not flow unless it contains significant amounts of water. Remember that freshly cut wood contains almost half its weight in water. Even in perfectly dry wood, up to a quarter pound of water is formed from the combustion of one pound of wood.

How to Avoid Creosote Buildup

Burn Hardwood

The old timers say good dry hardwood is the best to burn, the heavier the better, and in actual practice, this turns out to be true. Almost without exception, hardwoods contain about 8600 Btu per pound. One would then presumably choose to burn fewer pieces of heavier wood to gain the same amount of heat. Softwoods contain about as much heat per pound and in some cases more, but they weigh much less generally than hardwoods. Pitchy softwoods contain much resin and gum. These substances are high in energy content but converting their potential into heat is difficult. They also increase tar fog emissions and, eventually, creosote buildup.

Burn Dry Wood

Water content of the wood you burn has a lot to do with creosote buildup. Dry wood, of course, is better—mostly because it can be burned with less draft. Moderately dry wood is better from a creosote point of view than very dry wood. Very dry wood burns quickly and hot, so for a given desired temperature (assuming the desired temperature is less than the maximum stove output potential), the stove draft must be closed down to inhibit too rapid burning. This lesser draft means higher smoke density, longer residence time and lower stack temperatures—all of which contribute to creosote buildup. In a fireplace, where the draft cannot be significantly affected, dry wood is the best bet. It's easier to get started and it puts out more heat per pound than green wood.

The familiar radiating cracks of seasoned hardwood distinguish it from unseasoned green wood at right.

Have a Short, Straight Stovepipe

The configuration of the stovepipe and chimney is also an important factor in reducing buildup. It should be as short and have as few bends as possible. (Again, to recover the most heat, this is undesirable.)

Superfluous elbows reduce effective draft and add turbulence, thus contributing to creosote buildup. Proper size pipe or flue size also is helpful. Too small will not provide proper draft; too large increases residence time and lowers overall stack temperatures.

Select the Proper Size Stove

The proper size stove will help you keep creosote buildup at a reasonable level. Many people buy stoves too large for the area they wish to heat, thinking overkill is better than not enough heat. Unfortunately, when a stove is too large, it must be run at the low end of its output range. This means low damper settings, slow smoldering burns, and fast creosote buildup. Better to have a smaller stove that can be run at higher temperatures. Most stove manufacturers recommend heating areas for each of their models, and it is wise to use them as guidelines, bearing in mind particular aspects of your installation that might require a slightly different size. For example, a 1500-square foot underground house has differ-

A small stove may be all you need if your room is small.

ent heating requirements than a 1500-square foot uninsulated farmhouse on a windy knoll.

Burn Small, Hot Fires

Build only the size fire you need. Don't make an inferno that must be controlled by shutting the draft. This will restrict oxygen, causing poor combustion, longer residence time and lower stack temperatures. It is better to build a small fire that will not have to be heavily damped to obtain the proper amount of heat.

Add Fuel Often

It is better to add fuel to the fire in small amounts and often than in large loads. The pieces should be fairly large, too, to reduce the initial amount of smoke.

Keep Your Chimney Warm

A warm chimney will collect creosote more slowly than a cold one. For this reason, chimneys that are contained within the house are more desirable than exterior chimneys. Insulated chimneys—either masonry chimneys insulated between liner and brickwork with vermiculite or prefabricated chimneys, such as *Metalbestos* brand—keep the flue gas hot until it leaves the chimney.

A warm chimney has other advantages, too. Inside chimneys do radiate the heat that they collect into the house instead of outside. They also make starting fires easier as the draft is already present, and the temperature of the chimney is much higher, shortening the warmup time when creosote formation is fastest. An inside chimney will last longer than an outside one, too, because the weathering is less. Inside chimneys absorb much less moisture, thus having a longer life.

By now, the reasons for cleaning your chimney are probably clear. But just in case, here are a few more.

Heat transfer efficiency in the stove is most important. Creosote and ash on the sides and top of the stove can drastically reduce its heating capacity.

The movable controls on some stoves can become clogged or jammed by creosote buildup. Cleaning the stove should help these controls move more easily, adding to the pleasure and convenience of woodstove operation.

Smoke path blocks can occur in cookstoves or stoves with restricted vertical chambers. In this kind of stove, creosote can build up and flake off, falling to the bottom of the chamber. As this layer builds up, it can slow and even stop circulation. Cleaning will restore proper air circulation and enhance heat recovery in this way.

WHEN TO CLEAN

The variables leading to creosote formation are so many and so complex that generalizations are almost meaningless. However, here are some guidelines that should help you decide when and how often to clean your fireplace, stove, or chimney.

Fireplaces

Fireplaces, generally, don't need cleaning every year unless they get heavy use—four or five fires a week, for example. Of course, the variables of fireplace size and construction, type of wood, moisture content of wood, type of fires and kind of chimney all come into play when determining how often to clean. The best way to see if the chimney needs cleaning is to look and see. To do this, get a flashlight and look at the upper part of the firebox, or preferably, up into the smoke rise box. If the buildup is ¼ inch or more, clean the fireplace and chimney. My advice is to inspect often, especially when conditions of burning change. Installation of glass doors or a new delivery of wood may change the rate of creosote buildup, for example.

Stoves

Generalizing about frequency of stove cleanings is also a bad idea, since the conditions of burning vary so much. Stoves should be cleaned once a year at the very minimum. Do it in the spring, if you disconnect and store

Inspection is the way to tell that your stovepipe needs cleaning—as this one does.

the stove over the summer. Clean more frequently if the stove is the primary source of heat, if the stove is large for the area it heats and is run at the low end of its performance range, or if inspection of the firebox shows any significant buildup. Remember, ¼ inch of soot equals a 10 percent efficiency drop.

Spring cleaning is a good policy. If your stove isn't too heavy, you may wish to remove it for the summer, and on your way to storage, you can clean the stove outside. This will save you a bit of work and a lot of worry. If the stove is to be stored where temperature changes are likely to occur, cleaning will reduce corrosion caused by moisture condensation. Ash is very corrosive when combined with water—it forms lye. This is also an effective degreasing agent for the same reason, so it is important to remove all ash and creosote before storage.

The most reliable method of determining if your stove needs cleaning is inspection. Looking inside the stove is the easiest way, but not the most reliable. One should look inside the *stovepipe*, as the rate of buildup in the stovepipe is always faster than in the stove. Remove the stovepipe and look in it. If there is ¼ inch or more, clean it. Inspect it very frequently until you become familiar enough to know how long to go between cleanings.

Any change in burning pattern should be accompanied by frequent inspections. A new stove should be inspected after two weeks of burning. If it looks all right, go another two weeks; if it still is all right, try a month. After a while, you'll get an idea of how often you'll need to clean it. But if conditions change—say you install a heat reclaimer in the stovepipe, or get some new wood, or another person habitually operates the stove—start the two-week inspection program again.

You may clean the stove and chimney for convenience. A dirty chimney can cause back puffing or smoking when the fuel door is opened.

Clean your stove and chimney anytime you wish to get those marginal Btu's out of the stove, or when there is a perceptible drop in performance. My aunt has a mint condition 100-year old cookstove she was ready to junk because the oven didn't get hot. After I cleaned it for her, I warned her that it would get much hotter than before, but still she burned three batches of biscuits before she got used to the restored high performance of her stove.

Cleaning Your Stove

The tools required for cleaning your stove are not difficult to find. Although not all the tools listed are mandatory, the more you gather before beginning the task, the smaller the chance of problems arising. Beg, borrow, or steal the following.

- drop cloth
- trouble light
- old pair of leather work gloves (this may be their final destiny)
- hand wire brush (I prefer the long-handled type with a brush area about 1×5 inches)
- hand scraper or stiff putty knife
- hammer and screw driver
- ash shovel and metal bucket
- vacuum cleaner (We hope you won't need this. Using your household vacuum should be avoided; rent an industrial one.)
- dust pan and whisk broom

A can of furnace or refractory cement and an adjustable wrench are good standbys if you must disassemble any parts of the stove.

Slowly is Best

When you have assembled the tools needed, take a moment mentally to prepare yourself for the task. Read the procedure outlined in this book

Tools you'll need: ash bucket and shovel, whisk broom, hand wire brush, drop cloth, trouble light, flue brush, and brush extension handles. (Not shown: scraper or stiff putty knife, hammer, screwdriver, dustpan, and old leather work gloves.)

through once before beginning. Wear clothes that you can get dirty. You must not be in a rush. Accept that you will get somewhat dirty but resolve to keep the house clean. Remind yourself that your misgivings about the task have no real basis. Thus morally fortified, proceed.

As in most chores, the notions of ease and expediency must be kept foremost in one's mind. Take care not to weigh the expediency too heavily or you may find the ease way out of balance. I say this because many a sweep trying to finish his job quickly and hurry home to dinner has wound up spending much more time cleaning up the mess his haste had caused. Ash is extremely light and susceptible to the whims of the air around it. It is very easy to spread fly ash dust without realizing it. The best prevention is to do everything slowly.

Clean a Cold Stove

Before beginning, be sure the stove is stone cold. Working on and in even a warm stove is possible, but highly undesirable. Imagine hot ash forming a uniform coating on your perspiration-soaked arms. If that thought doesn't convince you to wait until the stove is cold, imagine a hot coal buried in the ash, still glowing from a fire two days past, falling inside the cuff of your glove. If possible, it is often more convenient to take the stove outside to clean it.

Remove the Ashes

Begin by spreading the drop cloth. If the stove is too heavy to lift, two cloths are handy. Don your gloves and remove the ashes from the stove. Some stoves are equipped with ash drawers that make for easy removal of ash. Others must be shoveled out by reaching through the fuel door. Top-loading stoves are difficult to remove the ashes from.

If the stove does not have a grate, some protection for the base of the firebox is required. Some people use sand in the base of the firebox; others use a good bed of ashes. At any rate, do not remove *all* the ashes. If it is necessary to add sand when you finish the task, do not add sand intended for road use; it may contain salt. Salt and iron do not complement each other, particularly at high temperatures.

Carefully shoveling ashes into the bucket.

When shoveling ashes, do it very slowly and deliberately—ash is one of the few substances that does not conform to the laws of gravity. If you treat it as if it were explosive, you'll avoid spreading any in the house. It is best to keep the stove as closed as is convenient and still have room to work. This will keep inside the stove whatever ash does become airborne.

If you sprung for the rental vacuum listed in the tool section, you can use it while shoveling out the ash. *I do not recommend use of your own home vacuum*. Ash is acidic and will deteriorate the cotton that many filter bags are made of. The particles are so fine they may also enter the bearings of the motor, causing premature wear. If you do have the vacuum, use it as a *dust collector*. That is, use it to sweep the air clean of dust rather than to pick up quantities of ash. Put the hose in any opening in the stove that's convenient, or hold it in your hand so that you can "sweep" the air either in the stove or over the bucket that you're shoveling the ash into. Avoid sucking large amounts of ash into the vacuum.

Clean Inside Stove with Wire Brush

Many stoves are simple boxes with a door at one end and the smoke exiting a hole at the opposite end or top. If your stove is shaped like this, cleaning the inside is quite simple. When the ash has been removed, use

USE YOUR ASHES

Save whatever ashes you remove for use in the garden. Wood ash decreases soil acidity. It contains calcium, potassium, sodium, magnesium, iron, silicon, phosphorus, and sulphur. Ashes should be kept away from some plants and seeds and seedlings. By contrast, roses, the chimney sweep's flowers, love ashes.

Wood ashes can also be used to make soap. Very simple soap can be made from ash and lard. Mix ash with water for an effective degreasing agent for auto parts. If you use ash for this purpose, be sure to rinse the parts well. You can also remove hair or fur from skins using ashes.

Some people use ash on their walks to keep from slipping on ice. I find that it works, but a lot of it gets tracked into the house. It works very well for getting cars unstuck, though, and I carry some in a can in my car.

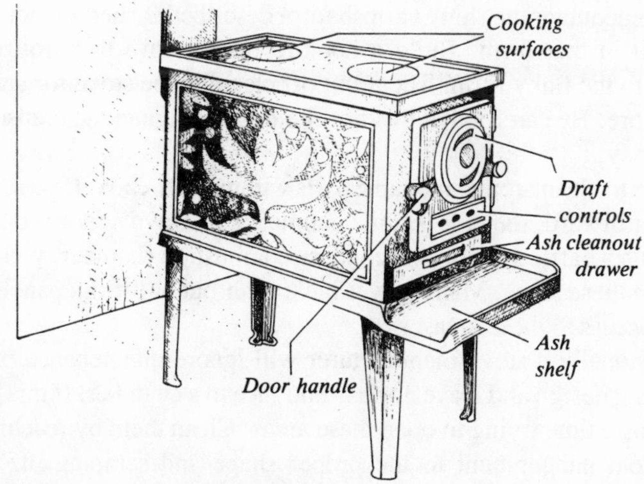

A simple box stove, showing basic parts.

the trouble light to illuminate the inside and peer in. Once you have an idea what the inside looks like, use your wire brush (gloves on) to brush or scrape every accessible square inch inside the stove. Do not try to restore the surfaces to a metallic color—just remove as much creosote as possible.

Brushing clean the entire stove inside—including cooking plates and all removable parts.

If you encounter the shiny varnish form described earlier, do not try to remove it in that form. Build a hot fire in the stove to pyrolize the creosote to the flaky form. Then, put off cleaning the stove for another day or more. Be careful that this hot fire does not ignite the creosote in the chimney.

Be sure to clean around movable parts within the firebox. If your stove is straightforward, the task is quite simple. If you own a more complex stove, with a baffle system or secondary combustion chamber, you must also clean these areas. Many stoves have push-out or lift-out panels that provide access to these areas.

Occasionally a stove manufacturer will ignore maintenance procedure during design and leave no easy entrance to a chamber. If this is so, do not waste time trying to open these areas. Clean them by reaching in with a coat hanger bent to the proper shape and scraping off what creosote you can.

Simple Stoves

An example of a stove with lift-out panels is the *Jotul 118* pictured here. The top of this stove simply lifts off. The top should be carefully wirebrushed, taking care *not* to brush the asbestos seal near the edges. If you're cleaning a stove with an enamel surface, take care not to chip the enamel. Actually, any piece of cast iron, especially old cast iron that has been through hundreds or even thousands of heat cycles, should be treated very carefully because it is often very brittle.

With the top of the *Jotul 118* removed, you will see the *baffle plate*. This plate is intended to lengthen the smoke path and give the stove more time to get the heat out of the flue gas. Reach in and lift this panel out. It rests on its side edges, supported by two removable panels hung on the inside walls of the stove. Remove and brush each of these plates, noting their position carefully for reassembly. There are few enough pieces so that even if you forget the way they go in, trial and error will eventually get them back in right.

When all of the pieces have been removed and brushed, inspect the stove. (The inspection procedure is described on pages 36-38.)

Replace all the pieces. Carefully replace the top, taking care not to damage the seal. When the whole thing is back together, congratulate yourself and do the only thing any self-respecting chimney sweep would do—have a glass of wine.

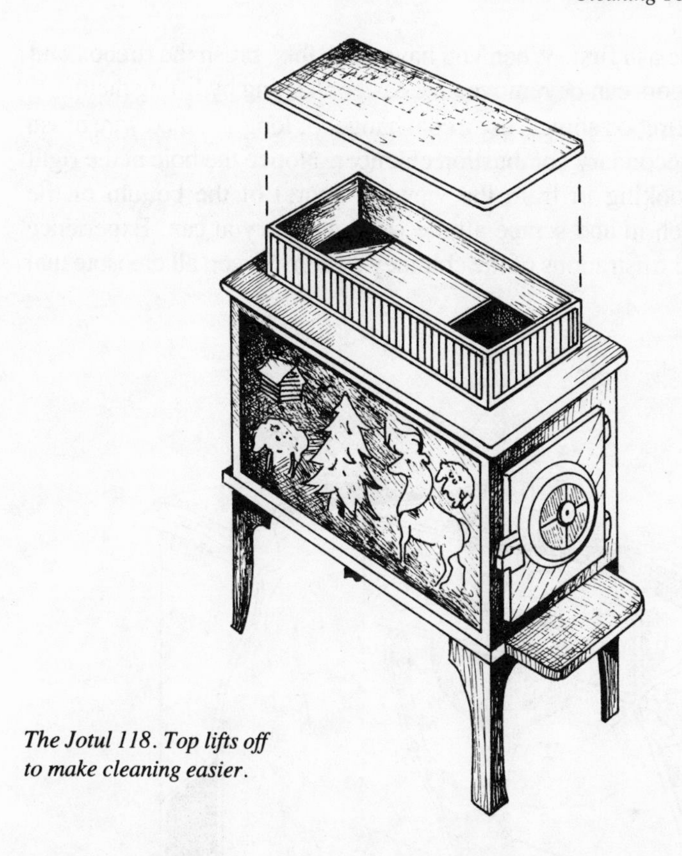

The Jotul 118. Top lifts off to make cleaning easier.

Complex Stoves

There are more complex stoves on the market today—the *Defiant,* made by Vermont Castings, is an example. To clean this stove, an example of inaccessible chamber–type stoves, first remove the ashes. This is best accomplished through the end or fuel door. The stove is equipped with an internal movable baffle. The handle has been positioned over the fuel door in such a manner that the door cannot be opened unless the baffle is open. This arrangement prevents the stove from smoking when the fuel door is open.

Open the baffle and open the fuel door. Shovel out the ashes. As you proceed, it may be easier to open the front fire viewing doors. *If you open them, beware of ash escaping*. Many stoves do not require removing all the ash to clean the stove. The *Defiant*, though, is easier to clean if you remove all the ash first. When you have done this, brush the firebox and doors. The doors can be removed for easier cleaning by lifting them.

When the firebox shines, use a coat hanger or long-handled wire brush to clean the secondary combustion chamber. Notice the hole at the right hand side (looking in from the viewing doors) of the bottom of the firebox. Reach in and scrape all the walls as best you can. Experience and enjoy the frustrations of the chimney sweep. Sweep all creosote that

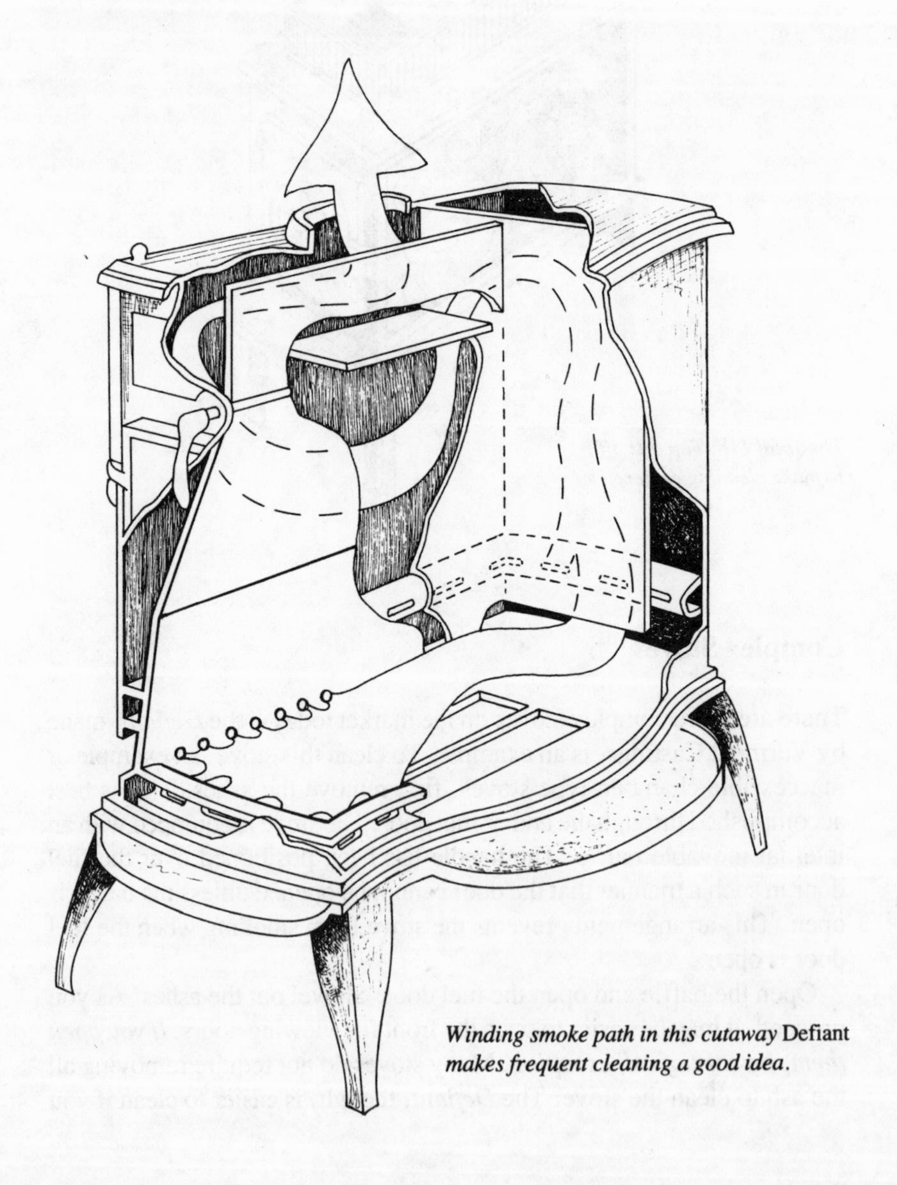

Winding smoke path in this cutaway Defiant *makes frequent cleaning a good idea.*

CLEANING COMPLEX STOVES—THE DEFIANT

1. From fuel door or front doors, shovel out ashes.

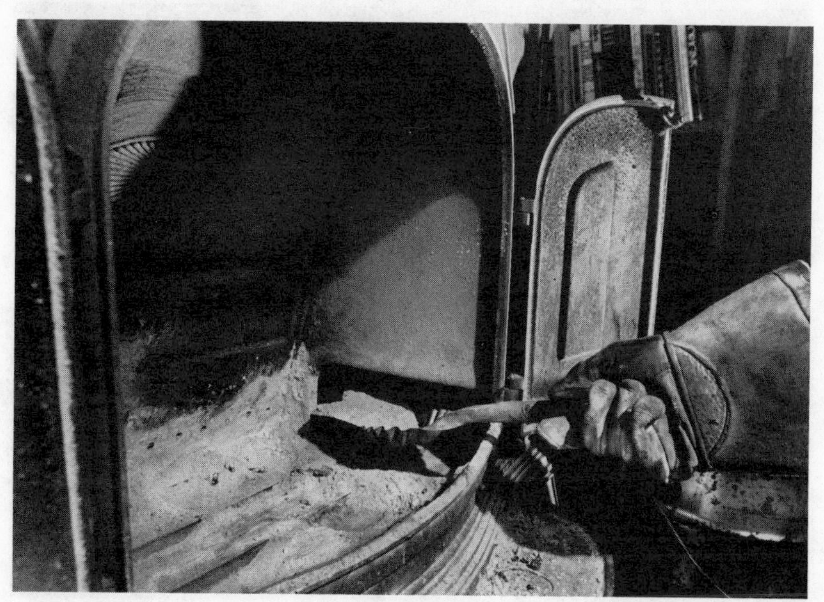

2. Be sure to shovel secondary combustion chamber.

3. Wirebrush this chamber carefully.

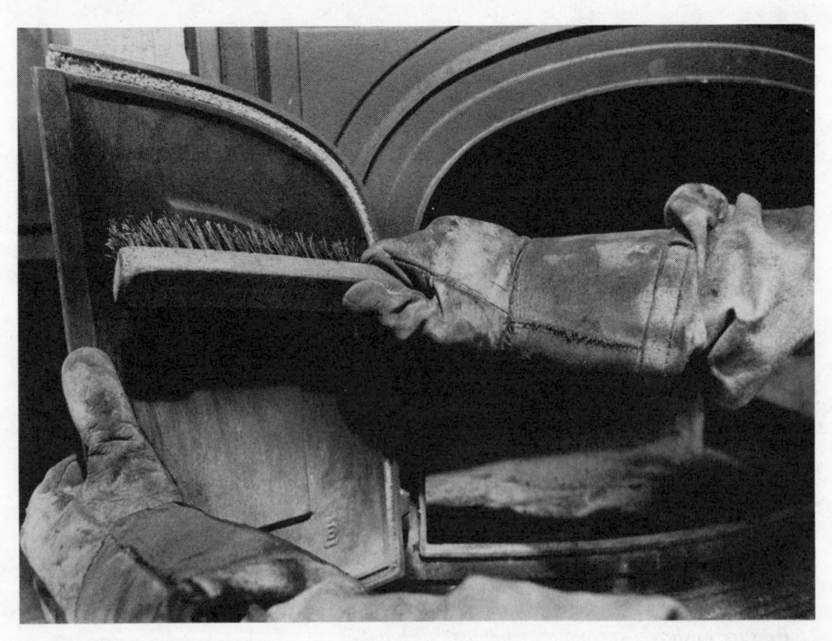

4. Brush the doors, too.

5. *Doors may be lifted off for easier cleaning.*

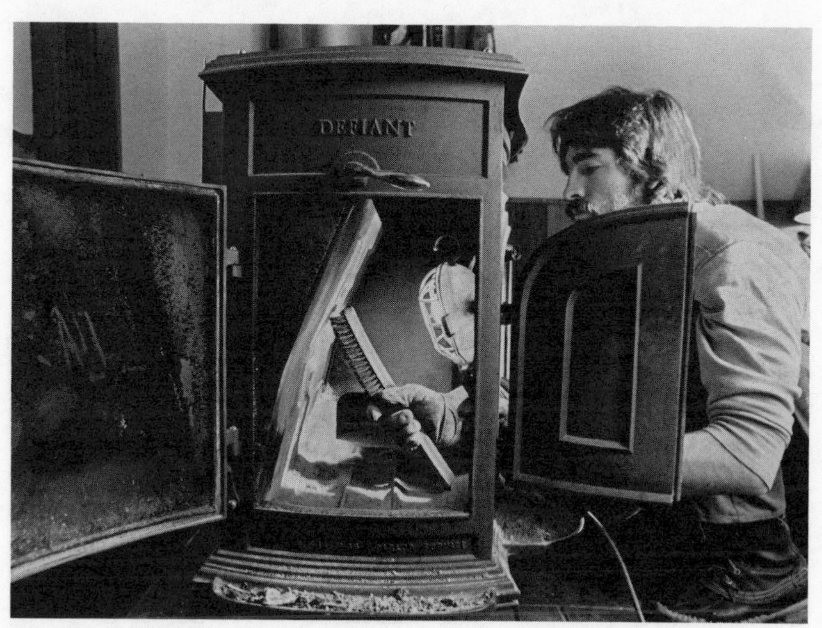

6. *Clean the baffle.*

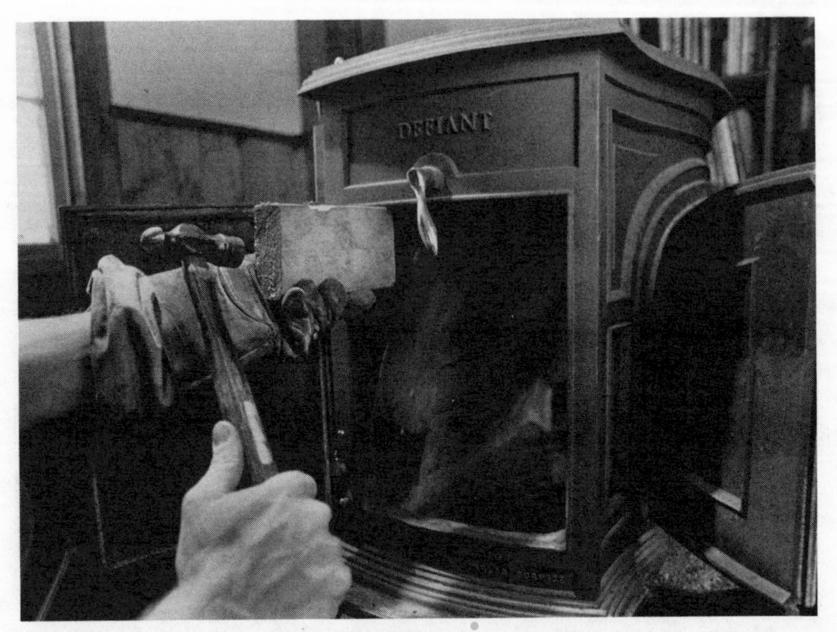

7. Padding the blow, you can remove the panel to clean behind on top exit stoves.

8. Lift off top plate from below. Brush it.

9. In brushing, respect the asbestos seals.

10. Remove the screws and disassemble the stovepipe.

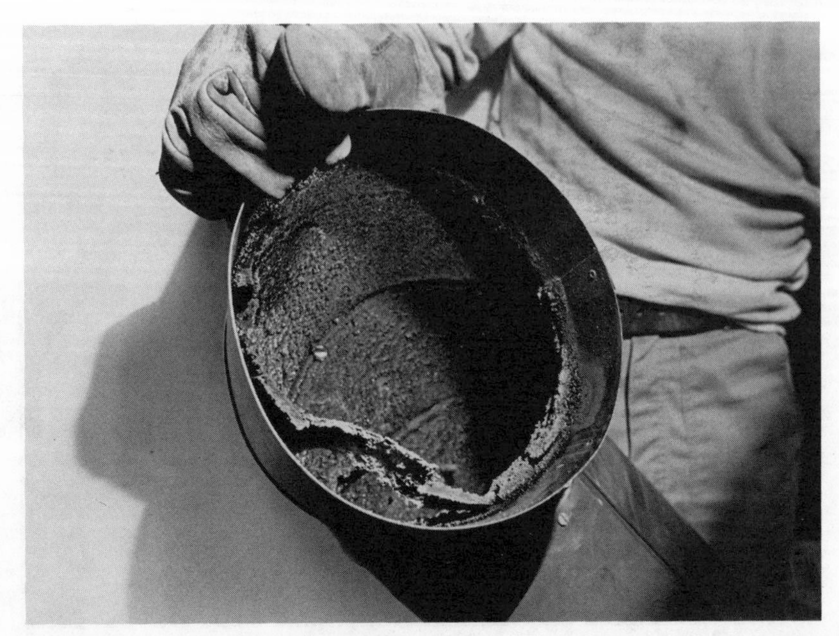

11. The elbows are the worst—brush them well.

12. The industrial vacuum may be the only tool that works in areas that are hard to reach.

13. When finished, replace a bed of ashes in the firebox.

falls down out of this chamber, as this is one place where creosote can block the smoke path.

The next chamber to clean is in the upper back of the stove. This area can be reached through the stovepipe hole. Reach in and clean this chamber. Notice the baffle at the front of this part. It opens into the firebox. Clean both sides of the baffle.

Close the baffle and look for the opening into the chamber below. Brush this one out. Cleaning this chamber is difficult. If you prefer, you can gain access to it another way. Open the end fuel door. There is a triangular panel in the upper left corner. This panel is designed to push out on top exit stoves only. Do not remove this panel on rear exit models, as it is impossible to replace. It is cemented in with furnace cement and may require a few taps with a mallet. *Don't* hit it with a steel hammer unless you pad the blow with a piece of wood. Do not remove this panel unless you have furnace cement to use when you replace the panel. Open the chamber and clean it. Brush the old cement off the seams. When you are ready to replace the panel, apply cement and insert the panel.

The manufacturer, Vermont Castings, no longer recommends removal of this plate by users, but now that you're a chimney sweep of sorts, you can take it out. Avoid it if you can. Just about the only way to get the creosote up from the crevice it falls into is to use the vacuum. Do not try to restore the shiny metal color to the inside, just brush off any perceptible buildup of creosote.

Check for Deterioration

When the stove is clean and apart is a good time to inspect it for signs of deterioration. Warpage of cast iron or mild steel can be expected, but can be tolerated only if the warps do not interfere with the function of the stove. Warpage in a welded stove usually is not a problem unless it prevents the door from sealing properly. Look for cracks in any of the castings or cracks that may have developed between castings.

When a cast iron stove is assembled at the factory, furnace cement is put on all the seams. With use or warpage, some of this cement may have disappeared. It may help to place the droplight in the stove, close the stove as much as possible, and darken the room. Look for tiny cracks of light. Use furnace cement to patch any holes. Generally, if the plates on either side of the crack are mechanically stable, cement can be successfully applied. A putty knife works to apply the cement, but nothing

GUIDELINES FOR SAFE STOVE INSTALLATION

The stove should be as close to the chimney as allowed by safe and practical procedure.

There should be a noncombustible shield or hearth under the stove that extends twelve inches on all sides of the stove and eighteen inches in front of the fuel door (see illustration). The National Fire Protection Association says brick, stone, or asbestos cement board do not serve as effective shields. They suggest asbestos mill board in combination with other materials, such as brick.

The stove should be a minimum of thirty-six inches away from combustibles. This distance may be reduced by placing ¼-inch asbestos millboard or sheet metal on noncombustible spacers one inch out from the wall. The thirty-six-inch requirement can be reduced to eighteen inches in this manner (see illustration).

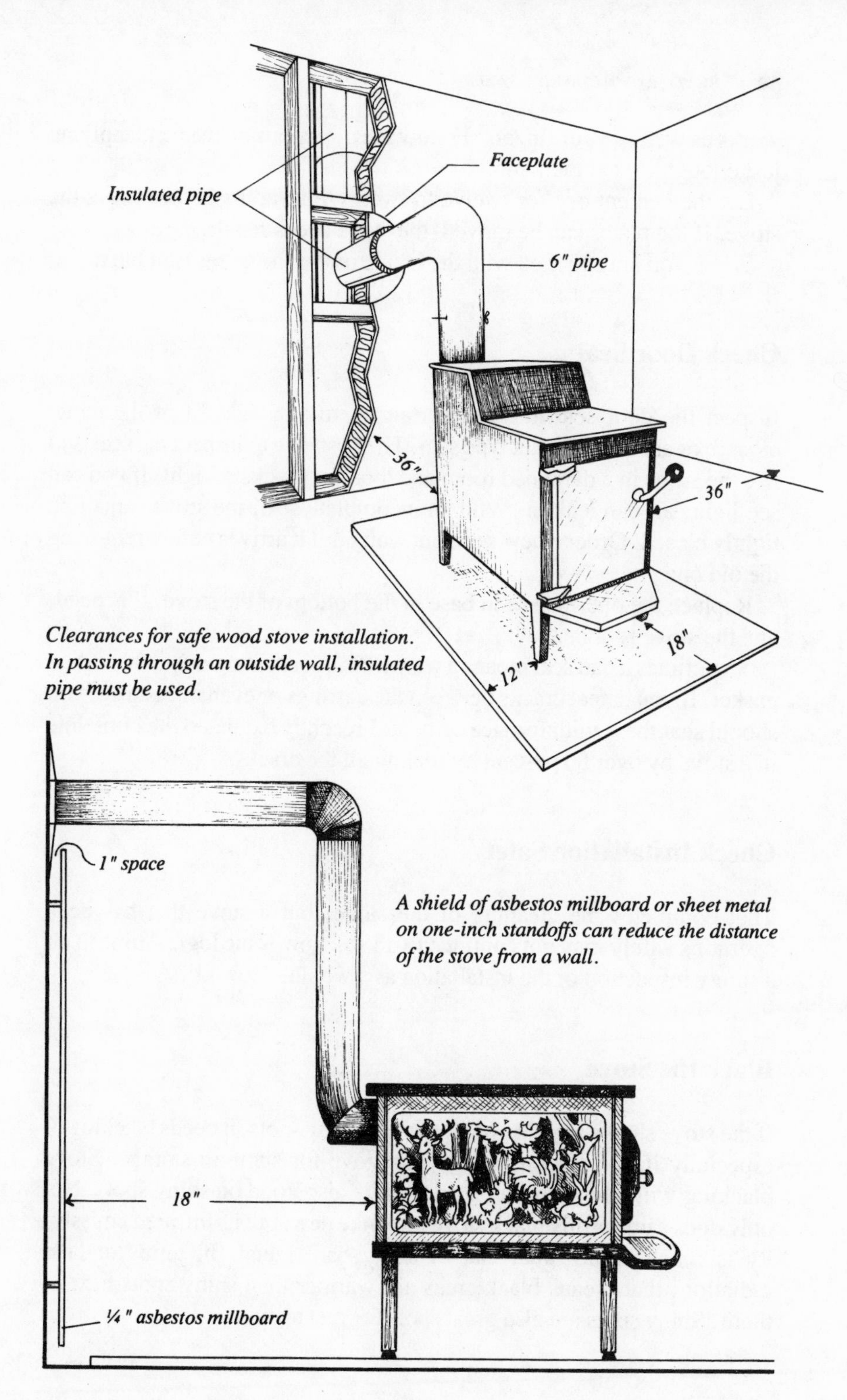

Clearances for safe wood stove installation. In passing through an outside wall, insulated pipe must be used.

A shield of asbestos millboard or sheet metal on one-inch standoffs can reduce the distance of the stove from a wall.

works as well as your fingers. Follow the directions on the can, applying just enough to seal the joint.

Let the cement dry for twelve to twenty-four hours before firing the stove. If the plates can be moved, the stove needs repair.

When you are satisfied with the inside of the stove, reassemble it.

Check Door Seal

Inspect the door seal to make certain there is a tight fit of the metal closure or asbestos-gasket type seal. The best way to inspect the seal is to fire the stove in a darkened room and look for cracks of light. If you can see light, air can leak in. With an incomplete seal, the stove cannot be tightly closed. Order a new seal, but wait until it arrives before removing the old one.

Replace the one-inch sand base in the bottom of the stove, if needed, and the stove is done.

Sometimes a bad seal means a warped door or casting rather than a bad gasket. If you detect cracks between the castings or even in castings, you should seal these with furnace cement. I recently extended the burn time in a stove by over 50 percent by sealing all the cracks.

Check Installation Safety

That completes the cleaning of the stove, but a stove that has been operating safely may not continue to do so. Now is the logical time to do a safety inspection of the installation as a whole.

Black the Stove

If the stove shows any sign of rust or any shiny spots, it needs blacking—especially if you are preparing your stove for summer storage. Stove blacking will inhibit further rusting. It is also good on shiny spots. Not only does a uniformly black stove look like new, but its infrared emissivity is much higher than that of shiny bare metal. In terms of heat radiation, that means black areas are warmer than shiny spots next to them. Shiny spots are also great spots for rust to start.

Cleaning a Cook Stove

Cleaning a cook stove is similar to cleaning a heating stove, with a few exceptions. Heat transfer efficiency is of great importance, so keep your cook stove clean. Again, do not attempt to clean the stove while it is hot or warm.

Collect the same tools as for the heating stove and lay out the drop cloth. Remove all the parts that can be lifted from the cooking surface and, with gloves on, wire brush them.

Shovel out the ashes or remove the ash drawer. Save those ashes for the garden, as they may be used to decrease the soil acidity and add phosphorous and potash.

Brush Stove and Baffles

Once you have laid open the bowels of the stove, methodically brush or scrape the entire inside. Determine the route of smoke passage from the firebox to the stovepipe and scrub the entire passageway. Observe the baffles that change the smoke path and direct the smoke and heat to different parts of the stove. Clean each baffle carefully to restore smooth motion.

Clean Oven Area

The oven area needs special attention. Baking in a woodstove is an art that requires slow, even heat. Many a wood-baked cake has been burned on one side while still batter on the other. Take great care to clean the

Top plates from cook stoves can be removed for brushing.

heating surfaces of the oven. Some stoves are equipped with special ports for this purpose; others require more patience. Either way, if you bake, keep the oven area clean. Remove the soot and creosote and replace the lift-out pieces.

Usually the smoke path in a cook stove leads from the firebox up and over the oven where it heats the top cooking surface. From there, it typically passes down the opposite wall of the stove, under the oven bottom, back to the back wall of the stove and into the stovepipe. See the illustration for this complex smoke path.

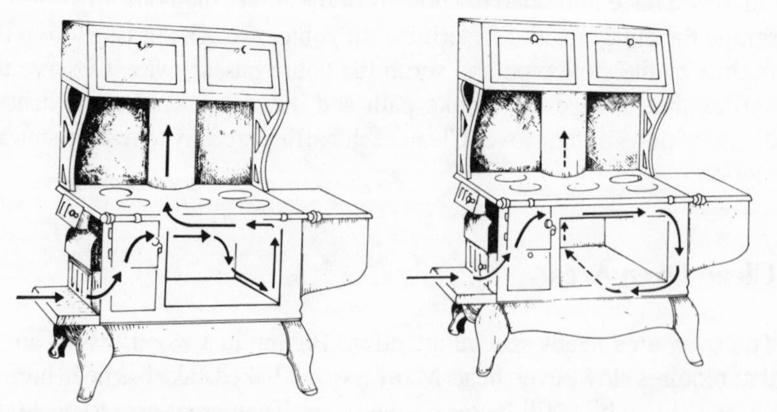

Smoke paths in cook stoves.

Sometimes the chamber under the oven is eliminated. The smoke path would typically then be down one side of the opposite wall and up the other side of the same wall. These narrow chambers collect thick deposits of creosote that impair the function of the stove in two ways: in the vertical chambers, creosote flakes off the walls and falls to the bottom. Eventually, it blocks the opening, causes the stove to smoke and eliminates air flow and heating ability of the oven. The creosote also cuts heat transfer efficiency.

Cleaning the Water Jacket or Coil

If your stove has a water jacket or coil that is in use, occasional cleaning of the water side is recommended. How often cleaning is needed depends on the mineral content of the water, the volume of water that passes through the jacket and whether a water softener is used. Biannual cleaning should suffice unless you have excessive minerals in the water.

If your stove is equipped to heat water, by all means do so. Heating water consumes valuable energy, and for a small investment, you can at least preheat the water entering your hot water heater and benefit by lowering your gas or electric bill and increasing your supply of hot water.

Cleaning of the coil is best done by a strong solvent that removes scale and muck from the inside of the tubing. In the past strong corrosives such as sulphuric and hydrochloric acids were used, but they were dangerous to handle and dispose of and they corroded the metal.

The best solution to use is a product called *Nutek 500* that is not acidic until heated and is biodegradable. We hesitate to say that it was developed for the cleaning of nuclear reactor watersides, but it remains the best product. It is probably not carried by your local stove store, but it's worth shopping for. Try a commercial boiler supply house or a boiler maintenance service. Don't let some slippery-tongued boiler salesman sell you acid. Specify *Nutek 500*.

You will also need a short piece of flexible tubing with a fitting that matches the upper connection of the coil, two buckets, and an end plug that matches the lower fitting of the coil. These fittings are outside the stove.

When you have collected these things, begin by closing the valves that lead to and from the domestic water supply, then uncoupling the top fitting of the coil and attaching the section of flexible hose. Fill the hose with water, and put the end in a bucket of *Nutek 500* solution. Raise the

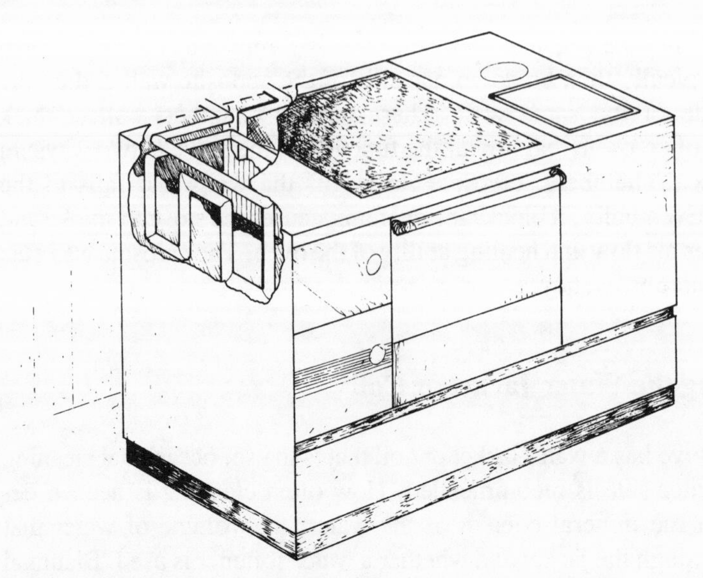

Wood cook stove with water-heating coil.

bucket high enough to set up a siphon and disconnect the lower coupling. You want the water to drain out of the coil into the second bucket, sucking the *Nutek 500* into the coil.

When the coil is filled with solution, seal the lower coupling with the end plug. Light a fire in the stove. The fire is necessary because the *Nutek 500* does not act as a cleaning agent until heated. Maintain a fire just hot enough to boil the solution for four to six hours, remembering that expansion and boiling will push the now-acidic solution out of the upper coupling and back into the bucket. If you use plastic pipe or hose, *beware of overheating* the plastic. If the plastic gets too soft, wrap a cold wet towel around the hose and connection.

At the end of the boiling period, remove the plug from the lower fitting and drain the solution, replacing with fresh water via the siphon technique. Heat this water to boiling and redrain, replacing with fresh water. Continue flushing with *cold water* until you feel comfortable reconnecting the domestic water (you're actually safe after the second flush). Despite the biodegradable claim, take the used solution outside and dump it where it can't contaminate open water.

Cleaning the Stovepipe

The stovepipe leads from the stove to the chimney or stack. It is sometimes called a *chimney connector* or *smokepipe*. Again, how frequently it should be cleaned can be determined by inspection. Stovepipe that carries low-velocity, high-concentration, relatively cool flue gas from a high-efficiency stove may need brushing out every other week, while stovepipe carrying high-velocity, hot flue gas could go an entire season before needing cleaning. Look for ¼-inch buildup; if you find it, clean the smokepipe and chimney. If it is less than ¼ inch, but is the shiny kind of creosote, burn the stove hotter (beware of setting a chimney fire off with higher heats).

When to Clean

Although cleaning the stovepipe has been left until now, in some installations it is advisable to clean the stovepipe before cleaning the stove. My house, for example, has a straight stack leading into a straight stovepipe. I sweep my chimney from the roof, pushing the brush right down the stovepipe and almost into the stove. That way everything falls into the stove, and I don't have to take anything apart.

The Flue Brush

Stovepipe can be cleaned with a hand wire brush, but it must be taken all apart to do this. A better solution is to purchase a flue brush for this purpose (see photo, page 22). The brush needs a handle, and, if your stovepipe is long, you can buy many extension handles that screw into one another to any length desired. Buy a brush that is sized to your pipe: a six-inch stovepipe requires a six-inch brush. They are available in basine, plastic and steel. Buy a steel one. It will be slightly more expensive, but it will do a better job in less time.

Some have argued for plastic and natural fiber brushes, cn the basis that steel brushes cause fine scratches on the inside of the pipe that might lead to rapid creosote accumulation. We feel this notion has little merit, since our experience has not shown this to be true.

Steel brushes are available with both round and flat bristle wire, and both kinds work well. Some people prefer the round wire type because the open area between the individual wires is larger and allows creosote and soot to fall through the brush. Clogging is not usually a problem, but it may occur on long runs. The brave soul who cleans his chimney while the stove is burning will find that smoke will pass through the round wire brush more easily. The flat wire brushes are prettier and slightly more expensive.

Brushes are available in about twenty-five standard sizes ranging from five-inch round to sixteen-by-twelve rectangular. Any size in between can be made at a slightly higher charge. Expect to pay anywhere from about $13.00 for a six-inch round to $35.00 for a sixteen-by-twelve-inch round wire brush. Flat wire brushes will be slightly more expensive.

Rectangular and square brushes are intended for tile flue liners. Liners are not uniform in size. Each batch of tile has slightly different composition that reacts differently when fired. In addition, different manufacturers make slightly different dimensions, and different parts of the country use different standards.

A stovepipe ripe for brushing. Right, the same pipe ready to be reassembled.

The flue brush should fit the pipe snugly.

To find what size brush you need, it is best actually to measure the flue opening. Buy a brush that fits it exactly if possible, or the next size up if none will fit precisely. A flue measuring 6¾ inches by 10½ inches would take a 7-inch by 11-inch brush, for example. When you have the proper size brush, proceed with the stovepipe.

Handles

Buy enough extension handles to reach the entire length of your stovepipe. If your chimney is suitable for cleaning with your brush, buy enough handles to reach the top of the chimney. Handles come in a variety of sizes and weights. They can be twisted wire, wood, or fiberglass. For occasional use, 0.35-inch diameter fiber glass rods with pipe thread fittings are perfect. Professional sweeps prefer fiberglass rods up to 0.48 inch thick for their stiffness, but the added expense makes them not worthwhile for most do-it-yourselfers.

Length of the rods is not critical—shorter rods simply mean more rods and more connections. Shorter rods do screw together more easily and store more easily than long ones. Three-foot rods are good to look for.

A brush and rods enough for your chimney could run $60.00, so you might ask your neighbor if she's interested in sharing ownership. When you have your brush and handles, spread the dropcloth and go to work.

Disassemble the Pipe

The stovepipe sometimes can be cleaned in place, but it's usually easier to take the whole thing outside. To do this, disassemble the pipe into sections that can be conveniently carried outside, but first make a scratch at all the joints that will be taken apart and on all articulations of swivel type elbows. If there will be many pieces, number the sections. When you're trying to rematch screw holes and maintain visual plumb during reasssembly, the scratches and numbers will pay off. Do not unfasten every joint, but disassemble the pipe to pieces small enough to avoid touching anything on the way out, as this inevitably causes soot to fall on a white carpet. Carry the sections with the bends or elbows opening end up to reduce further the possibility of soot fallout. When all of the pieces have completed the treacherous cross-carpet journey and are safely outside, disassemble to sections that have no more than one bend to make for easy cleaning.

Brush It Out

Use the flue brush to scrub the creosote off the surface. Don't expect the pipe to shine; just get it to the point where no thickness of deposit is discernible. Brush out the entire length of stovepipe.

SUPPLIERS OF CHIMNEY CLEANING EQUIPMENT

Black Magic Chimney Sweeps
Box 977
Stowe, Vermont 05672

Hearth Enterprises
508 Shorter Ave.
Rome, Georgia 30161

Worcester Brush
Box 658
Worcester, Massachusetts 01601

Ace Brush
30 Henry St.
Brooklyn, New York 11201

Remove Stovepipe Devices

If the stovepipe has a downdraft equalizer or a heat reclaimer, it too will probably have to be disassembled. A *downdraft equalizer* is a device inserted into the stovepipe just above the stove. It is supposed to eliminate downdrafts.

A *heat reclaimer* is another device that can be installed in stovepipe. It is a heat exchanger that extracts more heat from the smoke. Commonly it will have a number of tubes running through the smoke through which room air is forced by a small fan (see illustration).

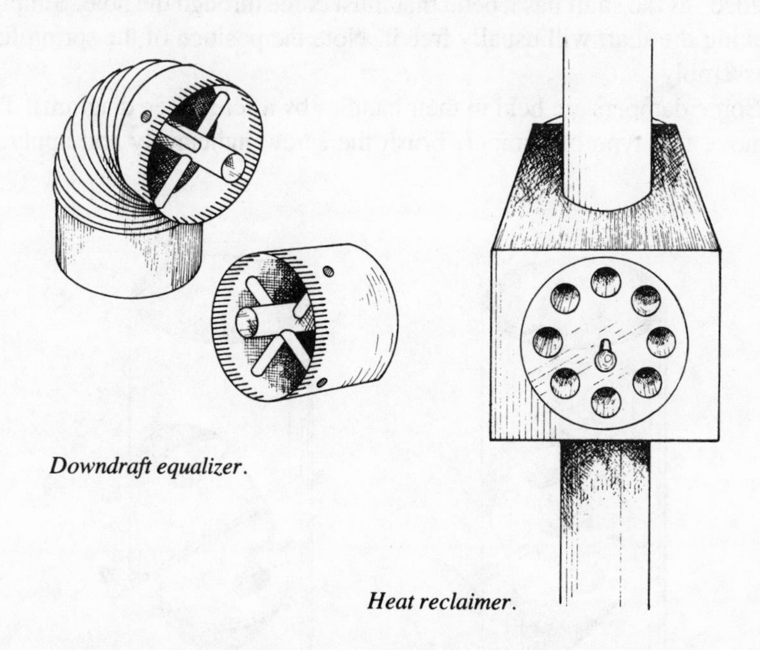

Downdraft equalizer.

Heat reclaimer.

Scratch the joint and take it apart. Brush out the equalizer if necessary. Most heat reclaimers have a cleaning plate that operates while the unit is in place. This cleaning plate often will jam because of excessive or particularly tough creosote. Scrub the heat transfer pipes until the cleaning rake will move again.

If you encounter creosote in the varnish form, don't try to scrape it all out. Reassemble the system and fire the stove hot enough to pyrolize the creosote to the more manageable, crusty stuff. The varnishy kind of creosote is an indication that you are running your stove too cool. Heat reclaimers contribute to this buildup by extracting even more heat from flue gases.

Remove Damper

The damper in the stovepipe is another potential problem. To remove it, reach inside the pipe with one hand and hold the damper plate. With the other hand, push the damper handle in, toward the pipe, and twist it a quarter turn. (See illustration.) This will disengage the handle and shaft. Simply pull the handle straight out of the hole. Some jiggling may be needed, as the shaft has a bend that must come through the hole. Simply rocking the shaft will usually free it. Note the position of the spring for reassembly.

Some dampers are held to their handles by a screw (see diagram). To remove this type of damper, brush the screw judiciously and apply a

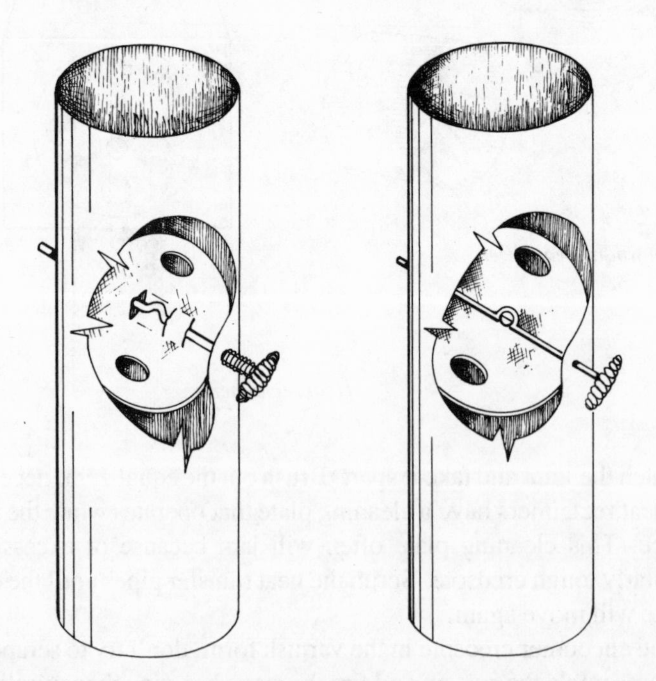

Two types of dampers. Remove the left one by pushing handle in and turning a quarter turn. The right one has a screw that must be unscrewed.

penetrating solvent like *Liquid Wrench*. High temperatures may have corroded the screw and one thing you don't need is a broken screw.

Some dampers are welded in and cannot be removed. If yours is this type, do not waste time trying to remove it. Brush it in place. Clean everything and inspect for corrosion. Remember, water from rain and condensation combines with creosote to form acid that is corrosive to steel.

Check the Pipe

A good way to see what shape your pipe is in is to give it a squeeze with your fingers. Unless you're very strong, firm pressure with your thumb should not push the metal. Of course, the pipe may flatten, but check for yield right around your finger. If the pipe is corroded, replace it.

There are a number of types and gauges of stovepipe available—galvanized, bright chrome, stainless steel and blue oxide are a few. When you buy a new stovepipe, bear in mind that stovepipe can contribute significant portions of heat to the house. To do its part to the fullest, the stovepipe should be black, but many people prefer a different appearance. Stainless is beautiful, but its thermal conduction is not high, even though its price is. Galvanized pipe is similarly low in its emissivity, and has a zinc coating that may vaporize. Buy the heaviest gauge black pipe you can find, especially if you occasionally burn trash in your stove. Smoke from trash corrodes stovepipe fast, so heavy gauge is safer and will last longer. One-piece, crimped elbows are better than swivel-type if you are making a 90° bend—they don't have seams that can leak air.

ASSEMBLING STOVEPIPE

Arrange the sections of stovepipe so that the male ends of the pipe point down. This will keep liquids from running out of poorly sealed joints and onto the outside of the pipe. Since the stack develops suction, smoke will not leak from these joints, even if they are not sealed with furnace cement. Assembling the pipe this way may be difficult, since the convention for years has been male ends up, but it's worth your trouble.

When assembling new stovepipe or reassembling your old pipe, be sure to use three screws at each joint. One or two screws do not hold the pipe securely in position. An insecure or movable pipe joint is undesirable because it may actually vibrate to pieces during a chimney fire. Keeping the whole system airtight will help to keep a chimney fire under control, so each joint should be sealed with furnace cement.

Apply the cement to the male ends of each section of pipe before assembly. As each new section is screwed together, apply the cement. Scoop up a gob of furnace cement and reach inside the pipe to the joint—be careful of the screws. Apply the cement, pushing it into the joint (see illustration). (Read the cautions on the cement can; many warn against exposure to skin.) Since most stovepipe comes in two-foot sections, you'll be able to reach far enough to do one joint at a time. Don't screw the whole pipe together and *then* try to apply cement.

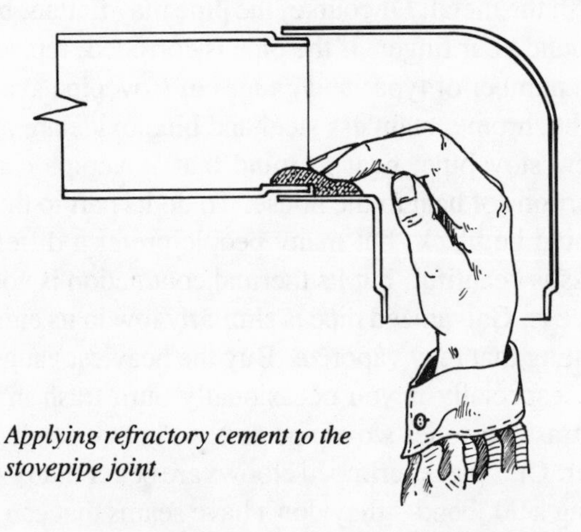

Applying refractory cement to the stovepipe joint.

Smooth the inside of each joint to allow uninterrupted air flow in the stovepipe. Doing the cement work from inside the pipe keeps the outside cleaner, but if you do get cement on the outside of the pipe, you can wipe it off with a damp cloth. (If you act fast.) Replace the damper and reassemble the pipe to the carrying stage, aligning the scratch marks.

Installing a Damper

If you're installing new pipe and wish to install a damper, get a damper that fits the pipe—eight-inch pipe requires an eight-inch damper. Drill a ¼-inch hole in the section of pipe that will contain the damper, making sure it will wind up at the height you wish. (To select the proper height,

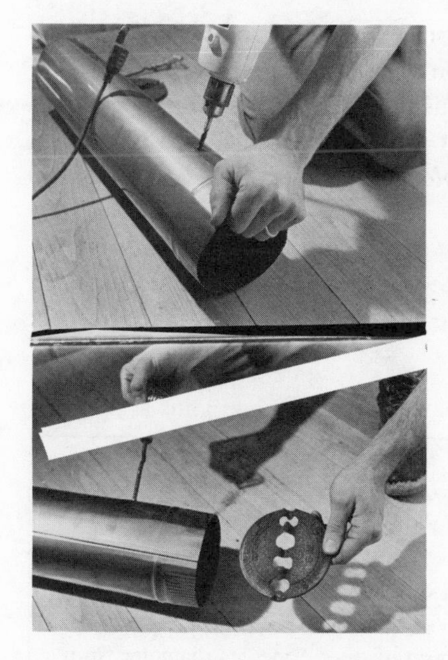

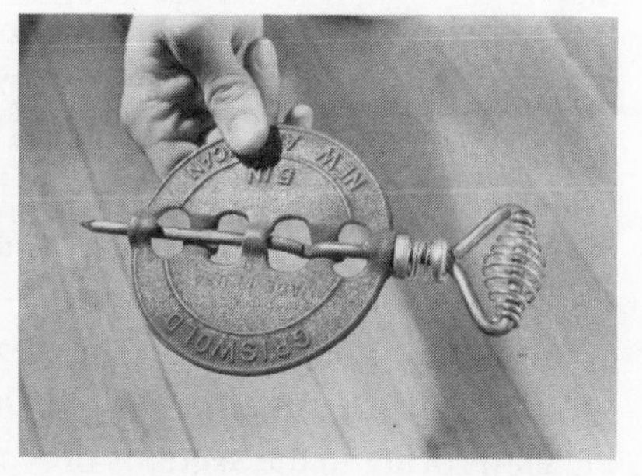

To install a damper, drill a ¼-inch hole near the end of one section of pipe (above left). The damper handle locks into position inside *the pipe (above right). Insert the damper plate and handle after drilling the second hole (left).*

make the handle easily reachable and far enough from the stove to keep the handle cool enough to touch.)

The position of the first hole is not critical if the pipe has not yet been screwed together. If it has, place the hole where you'd like the handle to end up. Hold the damper plate in one hand and insert it into the pipe as far as the hole. With the other hand, put the pointed end of the damper shaft through the hole in the pipe and fish around until you find the slot in the damper plate that the shaft is to slide into. Slide it in, turning the shaft relative to the plate, until the end of the shaft hits the far side of the pipe.

The location of the next hole is important, so proceed carefully. Feeling the plate inside the pipe, center it in the pipe with the damper in the closed position. See that the shaft is perpendicular to the pipe and push it against the far wall of the pipe hard enough to make a mark visible from the outside. Remove the damper shaft and the damper plate and drill another ¼-inch hole at the mark. Now install the damper.

Blacking the Stovepipe

If you are going to blacken the stovepipe, either for appearance or for more heat, do it now *before* you bring it into the house. Stoveblack of any type is very difficult to remove from carpets and sofas, so do the job outside. There are a number of products on the market for blacking stoves and stovepipes. They come as liquids, pastes and spray paints. I find spray paint very handy, but it must be paint intended for use on stoves. A product named *Thermolox 270* is very good. Do *not* use ordinary black paint.

If you are going to clean the rest of the chimney now too, leave the sections outside until the stack has been cleaned. If the chimney has been cleaned, bring in the sections of pipe and put them back in place. Inspect the safety of the installation.

GUIDELINES FOR
STOVEPIPE SAFETY

Here are some guidelines to insure a safe installation:

1. Stovepipe must be kept a minimum of three times its diameter from the closest combustible surface.

2. If the stovepipe must pass through a combustible partition, a correctly sized ventilated metal thimble can be used. The metal thimble should be three times the diameter of the stovepipe. Gas vent (which normally have very low stack temperatures and never have chimney fires) thimbles are not acceptable.

3. If stovepipe is used as a connector to the chimney, it should be cemented into the chimney with refractory cement. The connector should pass through the chimney wall to the inside surface but not beyond.

These safety guidelines may seem quite conservative, but they are designed to allow a margin of safety under any contingency. Twenty-four inch clearance between eight-inch stovepipe and the wall may seem excessive, for example, but in a chimney fire with the pipe red hot and radiating at temperatures in excess of 700°C. (1300°F.), twenty-four inches suddenly seems like just barely enough. Respect the three-times-the-diameter rule and sleep well at night.

Cleaning the Chimney

The chimney or stack is probably the most difficult part of the system to clean, but at the same time, the most important. This is where most of the flue gas condensation and resulting creosote occurs. It commonly has a much larger surface area than the stove or stovepipe, so it can accumulate much more creosote. A ¼-inch buildup in a chimney contains considerably more chimney fire fuel than ¼-inch buildup in a six-inch stovepipe. Although chimneys are tougher than stovepipe, they are also much, much more expensive and difficult to replace.

Condensation of the flue gas naturally occurs more readily on cold surfaces than warm, so if the stack is outside the house, buildup is likely to be faster than if it is inside the heated portion of the building. Often creosote deposits are heaviest near the point of smoke entry. This is a good place to inspect and determine the need for cleaning. Sometimes it is possible to clean only the stack without having to remove the stovepipe. If this is the case in your set-up, clean the chimney more often than the stove and stovepipe. How often can only be determined by inspection, so inspect often.

This section will include chimneys for stoves, for free-standing fireplaces, for wood or combination fuel furnaces and for boilers. Chimneys serving regular masonry fireplaces will be covered under the fireplace section.

CHIMNEY EFFICIENCY

A woodstove chimney has two functions. It carries the smoke from the fire out of the house, but it also provides the necessary suction on a woodstove to keep fresh air entering and feeding the fire. In a fireplace, this suction ensures that all the smoke from the fire is sucked into and up the chimney. The importance of keeping the stack temperatures up has been explained earlier, but little mention of the effect of chimney construction has been made thus far.

Several factors contribute to the effectiveness of any chimney: chimney diameter, height, material and resistance.

Chimney Diameter

Diameter has a large effect on chimney performance. As one can see from the accompanying table, increasing the chimney diameter from six inches to eight inches, for example, increases the relative capacity by a surprising 92 percent.

Chimney Diameter (Inches)	*Relative Capacity*
3	20
4	38
5	64
6	100
7	139
8	192
10	330
12	506

In this light, it is easy to see why small amounts of creosote buildup can cause much smoking from a stove.

Chimney Height

Height has an effect on function, but the effect of the added height alone is not large. Commonly, a small addition to the height of the chimney will have a more dramatic effect by extending the chimney beyond wind pressure influences surrounding it than by increasing the capacity of the chimney.

As indicated by the table, increasing the height of a ten-foot chimney to fifty feet does not even double its relative capacity.

Chimney Height (Feet)	*Relative Capacity*
6	60
8	69
10	76
15	88
20	100
30	115
50	135

Reprinted with permission from the 1975 Equipment Volume, *ASHRAE Handbook and Product Directory*.

Chimney Material

The material used in construction affects its function, primarily through the cooling of the flue gases. Masonry chimneys exposed to outside temperatures and elements may display considerably higher heat loss coefficients if they absorb water (if the water subsequently freezes, the heat loss is higher but its significance is dwarfed by the destructive action of frost in the chimney). Installing a well-drained chimney cap and applying masonry sealer to the exposed chimney will eliminate both problems.

Chimney Resistance

The elbows and other elements that contribute to the effective resistance of the chimney are also an important factor. Resistance is caused by friction of the gases with the chimney walls. You can reduce creosote buildup and improve chimney function by keeping the resistance of the chimney as low as possible. If you wish additional resistance, (to slow the burn rate of a not-too-tight stove for example), you can easily add a stovepipe damper. Resistance within the chimney comes from all the components of the chimney—even straight, clean pipe—but it is higher for elbows, tees, and expansion couplings.

The practice of adding all the component resistances to find the system resistance coefficient is not advocated as a useful way of finding out if your chimney works, but it is useful to know of the relative benefits of lowered resistance. Flue gas velocity can be computed using resistance coefficients and flue gas temperatures.

Feature	*Resistance Coefficient*
Round Elbow, 45 degrees	0.2–0.5
Round Elbow, 90 degrees	0.5–1.5
Tee, or 90 degree sharp elbow (mitered), or breeching (smoke pipe inserted into brick chimney)	1.0–4.0
Straight Pipe or Flue	
4'' diameter	0.1 per foot
6'' diameter	0.07 per foot
8'' diameter	0.05 per foot
Expansions (e.g. 5'' to 6'' stovepipe adapter)	0.05–0.15
Chimney Top	
Open	0.0
Spark screen	0.5
Rain and wind caps	0.5–3.0
Stovepipe Damper	
Open	negligible
Closed	5–20

Source: Adapted partially from Table 9, *ASHRAE Handbook and Product Directory, 1975, Equipment Volume*.

If you are planning to build a chimney, some of this material will be of interest and use, but most people are faced with a given situation that is neither practical nor desirable to change. If your goal is to keep creosote to a minimum, keeping the temperature within the chimney high is still the best advice.

The next table shows typical heat loss from different types of chimneys.

Type	*Heat Loss Coefficient*
Heavy steel, clay or iron pipe	1.2
Dark or weathered steel stove pipe	1.2
Shiny steel stovepipe	1.0
Masonry, unlined	1.0
Masonry, tile lined	1.0
Double wall metal, air space	0.5
Double wall, metal, insulation in between, stainless-steel inner	0.3
Triple wall metal, air ventilated (no insulation)	effectively greater than 1.2

Source: Reprinted with permission from the 1975 Equipment Volume, *ASHRAE Handbook and Product Directory*.

There are two basic types of stacks—prefabricated chimneys and masonry chimneys. The prefabricated chimneys are generally easier to clean, so we'll look at them first.

PREFABRICATED CHIMNEYS

Prefabricated chimneys can be made of double- or triple-walled pipe, sometimes insulated, sometimes not. Some prefabricated chimneys are made of aluminum. These are intended for gas vents and are not satisfactory for use with wood. Be sure that the chimney you wish to use is approved for "all fuel."

A common prefabricated chimney brand is *Metalbestos*. It is made of two concentric stainless pipes separated by a high temperature insulation. (The insulation is not, the manufacturer claims, asbestos, but they won't say what it is.) The sections have twist lock connectors (see illustration) and can be fitted with a wide variety of accessories to aid in constructing a good chimney. This type of chimney provides good insulation to the flue gas, keeping it warm.

Another type of prefabricated chimney has three concentric pipes separated by standoffs, but it is uninsulated. This type of chimney is

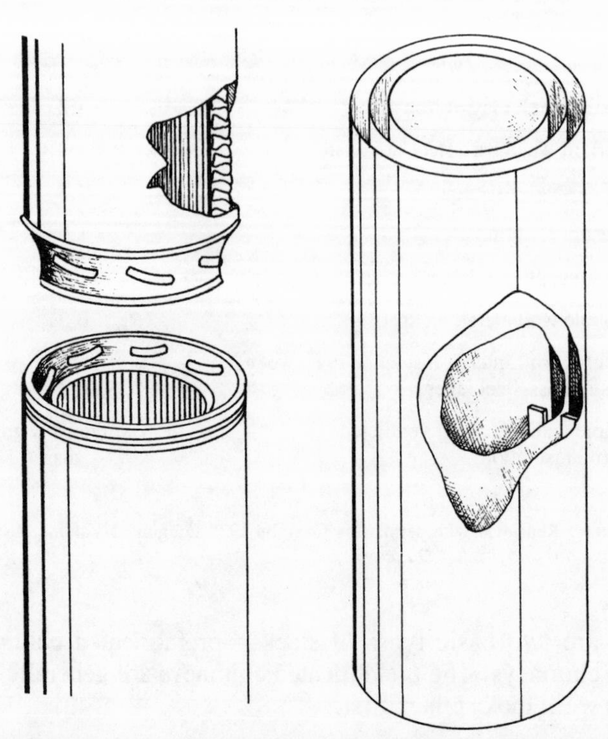

Insulated prefab chimney sections (left) twist together for a snug joint. Triple-wall pipe (right) is cooled by air. It builds up creosote fast and requires careful cleaning.

designed to be cooled by air circulation. The chamber between the inside pipe and the middle pipe contains air that is heated by the pipe carrying hot flue gas. It becomes bouyant and flows up the chimney and out the top. Air that rises out in this manner must be replaced from somewhere and since the bottom is sealed, it flows in from the top of the outside chamber. (See illustration.) This type of chimney cools the flue gas and builds up creosote relatively quickly. The downward draft in the outermost chamber also brings cold air into the house, causing an overall reduction in system efficiency.

Cleaning Prefabricated Chimneys from Outside

Cleaning either of these types of chimney is usually quite simple. The proper size fluebrush and handles, a screwdriver, a wire brush, the drop cloth and gloves are about all the tools required.

There are two different ways the chimney may be laid out. It may be mounted outside the building and enter through a wall or it may come straight down through the roof.

Wall exit with cleanout tee. This is the easiest type of chimney to clean. If the stovepipe is entering from the side, it is best to clean the horizontal section before the vertical part. If the stovepipe has not been removed, remove it. Spread the drop cloth under the opening left by the removed stovepipe. If the temperature is cold outside, so much the better. The bouyant warm air inside the house will want to rise up the chimney, carrying most of the dust and soot up with it.

Put your gloves on and screw the brush onto a handle and brush the horizontal section of chimney. When this has been done, remove the brush and pack the opening with crumpled newspaper or a rag. Take your brush and rods outside to the cleanout tee.

Spread the drop cloth under it if you don't want creosote left under the opening. Remove the end plug by prying it loose with the screw driver. Put on your gloves and go to work. Screw the brush onto a handle and

A cleanout tee like this helps to keep the mess outside the house. Just remove the plug and brush from below.

push it into the opening. Scrub the inside long enough to remove most of the creosote. Do not try to restore the original stainless steel gleam, just remove most of the creosote. Continue to add handles, scrubbing higher and higher up the flue until the brush reaches the top of the chimney.

Beware of pushing the chimney cap or bonnet off with the brush. Replacing the cap can take much more time than cleaning the chimney, especially if the chimney stands many feet above the roof. Likewise, when cleaning triple-wall chimneys from below, *be very careful not to push the top sections apart.* They are normally held in their standoff position by gravity. If they are pushed apart, getting them back together will mean going up on the roof, at the very least. More likely it means taking sections off the chimney, a time-consuming and wasteful task. It is usually safer to brush triple-wall pipe from above, where you can hold the pipe from sliding up and becoming disengaged.

Pull the brush back down from the chimney, removing the handles as they emerge from the stack. Replace the end plug, pick up the tools,

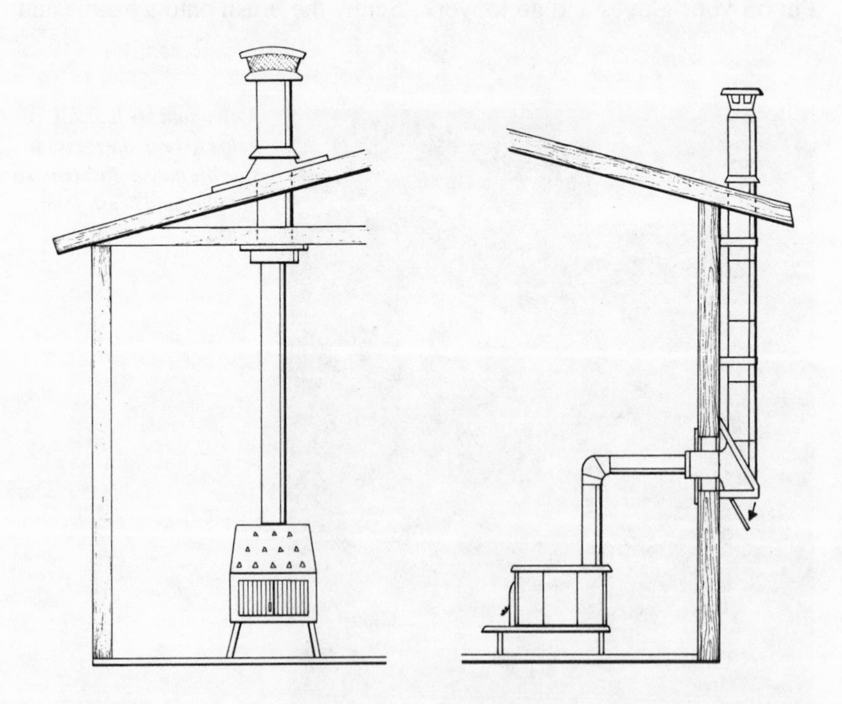

Two common prefabricated chimney installations. Left, a straight stack through the roof, an ideal installation. Right, an outside stack with cleanout tee—an easy one to clean.

gather up your cloth, and walk away. Before reconnecting the stovepipe, remember to remove the plug you placed in the horizontal section.

Roof exit with vertical stovepipe. The stack shown in the diagram can be cleaned either from below or from the roof. If the roof is not too high, too steep, or snowy, the easiest way to clean it may be by going onto the roof and brushing down. (Tips for going onto the roof are covered on page 71.) Assume you're standing beside the chimney.

Remove the *bonnet* or *chimney cap*. This cap is sometimes held in place with an expansion collar that fits inside the stack. It is expanded or contracted by a screw inside the *spark arrestor* (this is a screen of sorts on the outside of the cap). This screw may be hardly visible and it may be locked in place if there is appreciable creosote buildup. Caps sometimes are fastened with sheet metal screws on the outside of the stack.

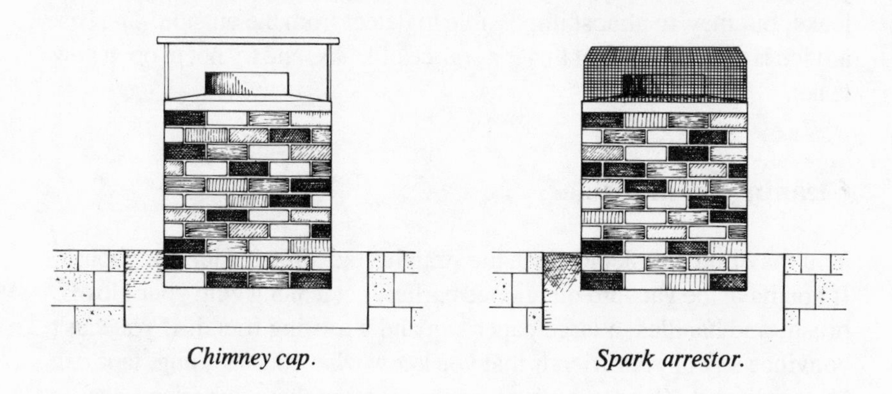

Chimney cap. *Spark arrestor.*

Once the cap is off, push the brush and rod into the stack. Brush that section until it is clean; add another rod and proceed to the next section. If you have triple-wall chimney, you may have to hold the inside pipe from popping up when the brush is drawn back up the chimney at each stroke.

If the buildup is thick, it is important to proceed slowly. As you knock creosote from the lining of the stack, much of it falls down. As it picks up speed, it causes the air in the stack also to begin descending. The higher the chimney, the faster the debris falls and the longer it has to act on the air column, making the effect more dramatic in tall chimneys. When you're on the roof cleaning, you can't see what's happening inside the house, so proceed as if dust is being pushed out into the house each time some debris falls.

After each downward stroke, pause long enough for the debris to fall to the bottom and the downward rush of air to cease. If you do this task when it is cold outside, a natural draft up the chimney will be established by the temperature differential inside and outside the house that will help carry away most of the light particles, keeping the house spotless. (A stove can sometimes be left burning while its chimney is cleaned in this manner, but the smoke and debris exiting the chimney top can make it a very unpleasant job. Falling debris could ignite and cause a chimney fire.)

Brush down the whole chimney (and if possible the stovepipe) all the way to the stove. Withdraw the brush and rods. Before replacing the cap, brush it with a wire brush. Be sure any spark arrestor is clean and won't cause additional resistance to the system. While you're on the roof, look for places that might leak water around the chimney. Any builder will tell you that where a chimney pierces the roof is a great potential source of leaks, but they're almost impossible to detect from the outside. The best advice is to be aware that this is a source of leaks, and try not to open new leaks.

Cleaning From Inside

You may find it easier to clean the straight stack from inside your house. If you have the vacuum mentioned earlier, get it along with your gloves, brush, and handles, a large paper bag and a trusting friend. If you can't convince any of your friends that you know what you are doing, tape can be substituted. (It may be best not to mention that your friend can be replaced by a roll of tape.) Spread the drop cloth under the stack and remove the stovepipe. Tear the paper bag partly down one side—the tear will allow the rods to be pushed into the stack, so make it the right size for that (see illustration).

Attach one handle to your brush and push it up into the stack just far enough to get it into the stack and hold itself. Have your friend hold the bag around the stack with the tear allowing the handle to extend out. If you were unable to solicit the help of a friend, tape the bag to the stack.

Brush the stack. If you use tape, be sure the weight of the debris collected in the bag does not pull the tape off. Brush slowly at first to avoid the downdraft often caused by falling creosote. Such a downdraft actually causes higher air pressure within the bag, pushing the dust that would normally be contained out through the tear. Brush particularly

CLEANING THE CHIMNEY FROM INSIDE

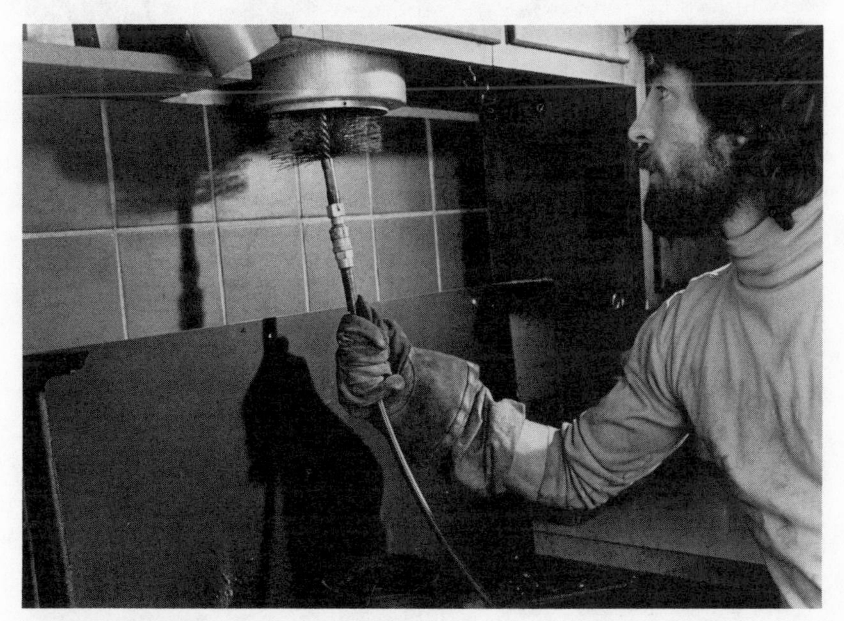

1. Insert brush with handle into chimney flue.

2. Using wide masking tape, fasten the torn bag to the chimney edge, working around the handle.

3. Make sure the tape is secure.

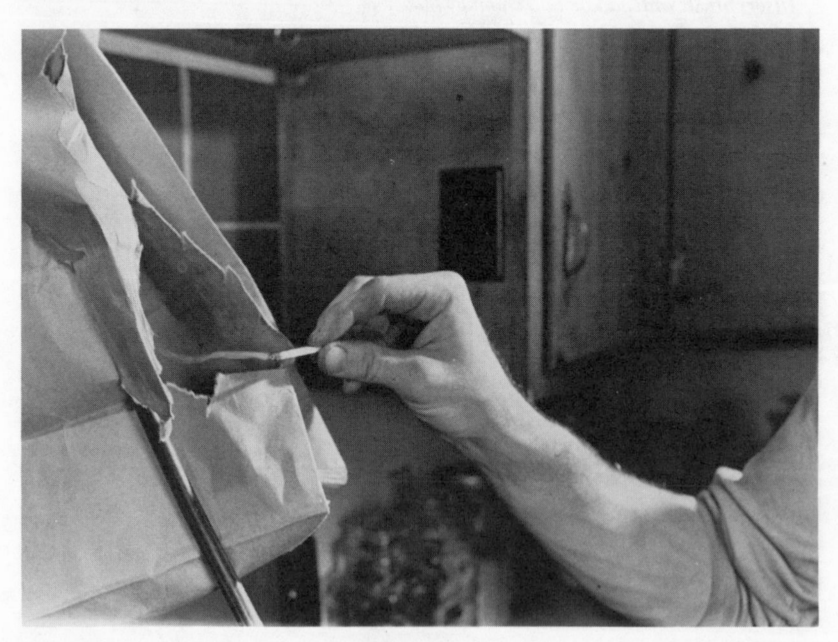

4. Test for a good updraft with a smoking match.

5. *After taping the bag closed, brush the chimney slowly.*

6. *When you finish, remove the bag and its cargo of creosote.*

slowly near the top of the stack, to avoid pushing off the cap or uncoupling triple-wall pipe. If your friend did his job particularly well, you may have no need for the vacuum. If not, the need will be self-evident.

Actual configurations of prefabricated chimneys may be slightly different from the ones outlined here, but the concepts for most installations are basically the same.

Inspect For Safety

When cleaning is complete, inspect the system for safety. Most prefabricated chimneys are approved for a two-inch clearance from combustibles. Check the manufacturer's recommendations and respect them. Two inches may seem overly cautious, but remember design is aimed at the worst possible situation—a raging chimney fire. See that all the joints are snug and properly engaged.

MASONRY CHIMNEYS

Masonry chimneys include those made of stone, brick, and block. They may be lined or unlined. Old masonry chimneys were made of stone laid up with mortar made of hydrated lime, clay, horsehair, and straw. In 1824, Joseph Aspidin invented Portland Cement and soon it replaced hydrated lime mortar as the best mortar. Ceramic flue tiles are a recent development, gaining popularity only within the past sixty years.

Flue Tiles

Most chimneys built today are lined with flue tiles. They can withstand much higher temperatures than mortar, and they provide a smooth path for the smoke to exit. Flue tiles come in more or less standard sizes and generally do not taper, making proper sized and shaped flue brushes ideal for cleaning them. Unlined chimneys usually can be cleaned with a brush of the proper dimensions, but their rough interiors and often varying sizes makes the task somewhat more difficult. Newer chimneys tend to be straighter than old ones. Naturally, the straight shafts are

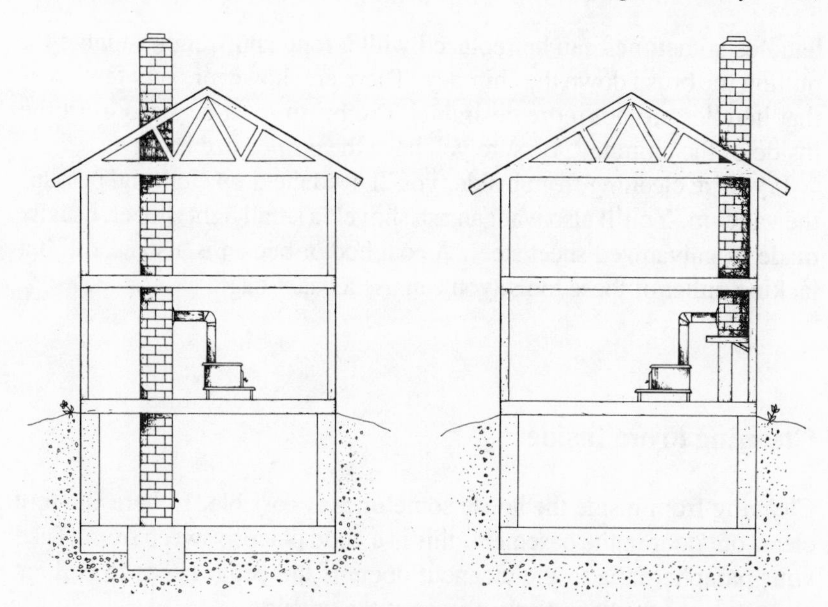

Masonry chimneys. Left, a full chimney has its own foundation and often an ash door. Right, a half-chimney with timber supports—creosote must be shoveled from the stove-pipe hole.

easier to clean, but you may have a chimney that makes a bend or two before it opens out to the sky. Usually these bends present little problem in cleaning. Cleaning around bends simply requires more time and patience.

Most chimneys begin at below ground level and extend through the roof (see illustration). Some, called *half-chimneys,* begin on wood supports built into the house. Such a chimney probably should not be built. If it is, it must be relatively short or the weight becomes prohibitive.

Cleaning either a half- or full chimney or a chimney serving a fireplace is almost the same. (We'll address fireplace chimneys in more detail in the fireplace section.)

Tools

The tools you'll need change slightly according to exact configuration, but you'll need at least a proper-sized brush and handles. The brush handles sometimes can be replaced with a rope and weight suitable for pulling the brush down the chimney. There are, however, very few times that handles are not more desirable. The brush must be sized to fit the inside of the chimney. Be sure you have the proper size brush.

If you're cleaning from inside, you'll need the drop cloth and perhaps the vacuum. You'll also want an ash shovel, a small light shovel, usually made of galvanized sheet steel. A coal hod or bucket is also useful, but lacking either of these tools, you can use a paper bag.

Cleaning From Inside

Cleaning from inside the house sometimes is possible. If there is a soot clean-out door in the basement, this is a good place to sweep from. Take your paraphernalia to the cleanout door. If the cellar needs to be kept clean, spread the drop cloth in front of the opening.

Check the draft in the chimney before proceeding. To do this, light a match and hold it in front of the open soot cleanout door. If the flame is

A soot cleanout door, this one outside. Because this door is close to the ground, it serves best as a place to shovel from after sweeping from the roof.

Sweeping a half-chimney from inside.

sucked in towards the chimney, the draft is favorable, so the majority of dust will be sucked up the chimney. If the draft is negative, or blowing out of the chimney, try to reverse the draft before proceeding.

A couple of things may be responsible for this downdraft. Any building whose temperature is warmer than the surrounding air generates what heating engineers call *stack effect*. The warmer and more bouyant air inside the house is trying to push upward. It exerts pressure on the roof and walls and leaks through. This air must be replaced and the easiest place for it to come from is often a chimney. Opening a downstairs window or two will often relieve the downdraft.

Another cause of downdrafts is wind. Opening a downstairs window on the windward side of the house often will eliminate downdrafts caused by wind. (If you cannot eliminate the downdraft, clean the chimney from the roof so the soot cleanout door may remain closed.)

Once the draft is moving upward, proceed by pushing the brush through the opening and up the chimney. The fiberglass rods are supple enough to allow quite a sharp bend. Sweep the chimney up and down a few strokes as you continue to add extension handles. Be sure to work slowly to avoid downdrafts caused by too much falling creosote.

When you reach the top of the chimney, you will feel a decrease in the resistance from the rods as the brush goes out the top of the chimney. Withdraw the brush, removing the handles as you go. When the brush is out, give it a few taps to dislodge whatever creosote still clings to it. This may save you some cleanup time. Use the ash shovel to remove the creosote from the bottom of the chimney. Close the soot cleanout door and pick up.

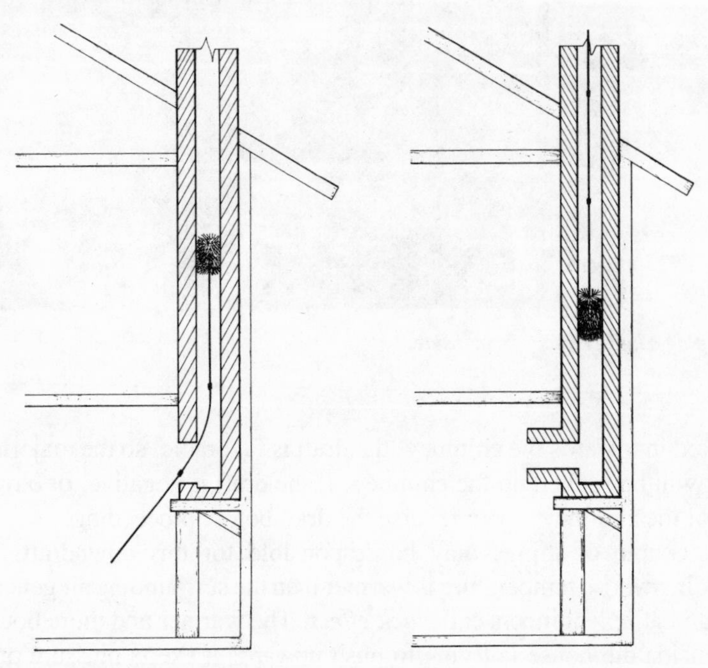

Cleaning half-chimney from inside, left, works fine if rods can make the bend. If not, right, you will have to clean from the roof.

The procedure for cleaning a half-chimney from inside is very similar. Often there is no soot cleanout door, nor even a place for the creosote to fall without partially blocking the chimney. If the chimney connector comes out of the chimney easily, the rods will be able to make the bend. If the connector is cemented in properly, you may have to clean from the roof, as the rods will not make too tight a bend (see illustrations).

If you can clean from inside, proceed as with the full chimney, taking care not to allow soot to flow out of the hole. When brushing is complete, slowly shovel out the debris. Replace the stovepipe, and you're done.

Cleaning From the Roof

Many chimneys are more difficult or even impossible to clean from inside the house. These you will have to clean from the roof. Many professional chimney sweeps prefer to sweep from the roof because of the different perspective of the world it provides. You may wish to experiment with roof-top sweeping just for the experience. If you do, try to obtain a top hat and wear it whenever you're in a place from which you would not wish to fall. Rumor has it that the hat will prevent you from falling off the roof. (My experience seems to bear this out—I have never known a sweep who fell from a roof while he was wearing his hat.)

Common sense will dictate the best way of getting on the roof, but there are some established guidelines. If you can get onto your roof and to the chimney without a ladder, good. Most people will have to use a ladder. If possible, position the ladder against a gable rather than against the eve of the roof, with at least one rung extending above the roof. Place the ladder on level ground so that it makes a four-to-one slope with both legs firmly planted. A four-to-one slope means that for every four units of vertical measurement, the base of the ladder should be placed one unit out from the wall (see illustration). If possible, have someone steady the ladder for you.

Before you climb to the roof, check to be sure that all possible places

For a ridge chimney, place the ladder on the gable with a 4:1 pitch.

A chimney in a sloping roof may be reached with two ladders, starting against the eve.

that dust could escape from the chimney have been sealed, and that you have the right size brush and enough rods.

If the stack comes through the roof at the ridge, you're all set. But if it emerges in the center of one side, you'll need another ladder and a ridge hook. Place the first ladder against the eve to enable easy placement and use of the second ladder. (See illustration.)

Once you are on the roof, the ridge or sometimes roof valleys are the easiest places to walk.

When walking on any roof, be careful not to damage it. On a dry roof, gum or crepe sole shoes are good. Sneakers also work well. Do *not* use leather-soled shoes. Soft shoes not only cling to the roof better, they also protect the roof from damage. Another tip is to keep as much of the sole of the shoe in contact with the roof as possible. The larger bearing surface will provide better boot "purchase."

If you must go on a roof with snow or ice, use boots that conform somewhat to the roof. Beware of snow on a roof suddenly sliding off—with you or onto you. Avoid frosty steel, slate, or wooden roofs like the plague. Frosty new steel roofs could possibly be the slipperiest thing known to man.

Slate roofs are particularly fragile, especially at the peak. Tile roofs should be treated like eggshells. Wooden roofs can be fairly tough, but if they are very old, they also are generally fragile. Avoid walking on asphalt shingle or half-lap roofs in very hot weather. Steel or tin roofs are usually quite tough, but they can be very slippery.

Reaching the chimney may require careful footwork.

Traditional sweeping is aided by modern brush handles, but the need for good balance remains the same.

Once you have reached the chimney, insert the brush into the top and sweep the chimney. Be sure that any ports into the chimney have been sealed or lead into a closed stove. Sweep slowly to avoid causing a downdraft.

Occasionally, a chimney has a cap large enough to prohibit the use of rods. If you encounter this setup, use a rope tied to the brush. Since you can't push the brush down the chimney with rope, you'll need someone to pull the brush down with another rope or a heavy weight to pull it down for you. Of course, if you use the weight, you'll also have to pull it back up.

CLEANING WITHOUT A BRUSH

There are several alternative methods of cleaning a flue without a brush, all of questionable effectiveness. The most popular second-rate method is the *hanging chain method*. This involves hanging a long, heavy chain down the chimney and depending on its weight and flexibility to knock off the creosote while it is swung and shaken from the top of the chimney. This method clearly cannot do the same kind of job the brush does. Worse than that, it can break flue tiles and damage the chimney.

Chains or *sand* in a burlap bag on the end of a rope are also poor methods.

A small evergreen drawn up and down the flue can be tried. In some households, it is traditional to select a small Christmas tree, and after New Year's, use the tree to clean the chimney. It often requires a great deal of strength to pull a tree down the chimney. (One farmer I know solved that problem by running a rope from the tree down the chimney, out the fireplace through the front door and to his tractor. The 'ole 8N didn't have a bit of trouble with that tree.)

You may prefer the *flapping goose method,* which involves buying your Christmas goose—alive—a day in advance. When you're ready to do in the goose, tie a rope around his feet and haul him up and down a few times. Remove the goose, decapitate and pluck. Then, of course, there is the *climbing boy* method—

Seriously, nothing works as well as a flue brush, so if you aren't going to buy a brush, consider hiring the services of a professional chimney sweep.

With a helper, you can clean by pulling the brush up and down the flue on a rope, eliminating the brush handles.

Cleanup

When the roof portion of the task is complete, you must remove the soot and creosote that have fallen to the bottom. If the chimney has a soot cleanout door at the base, shovel out the debris. If the chimney ends where the stovepipe plugs in, remove the soot there. Be certain to remove it all. A large amount of creosote will sometimes fall and pack itself in. One can shovel out all the visible creosote and still have a mass of it above, blocking the flue. To avoid this, poke something a foot or two up the flue once you think everything has been shoveled out. Put the debris in a plastic bag and send it to the dump.

MAKING A SMOKE TEST

When the entire system is reassembled and operational, make a smoke test. To do this, solicit the help of a friend. Send her onto the roof with a wet towel, ready to cover—and more important, ready to uncover—the top of the chimney. Go inside and light a very small fire in the stove. Add a handful of green leaves or hay or other material that smokes when it burns—though avoid rubber and other smelly materials. When you're getting good thick smoke, have your friend cover the chimney. Close the stove door, but keep your friend ready to remove the towel if too much smoke starts leaking into the house. Now inspect the entire chimney for leaks.

To make a smoke test, build a fire and cover the top and bottom of the chimney. Be sure that cloth is damp.

While you're doing it, inspect the chimney for other points of safety. This inspection is perhaps the most important one you'll make. The National Fire Protection Association *Handbook* (Section 7-57, 1976) states, "defective masonry chimneys are the leading causes of fires in buildings. Chimneys must always be made safe, even if it means rebuilding them." This statement is true for all fuels, but is particularly true with solid fuels, especially wood. Creosote from wood presents a much greater danger, for example, than does soot from an oil fire. Accordingly, chimneys connected to wood-burning stoves and fireplaces must be especially safe.

GUIDELINES
FOR CHIMNEY SAFETY

1. The chimney should have its entire weight on a foundation originating in the ground. It should be built on solid, frost-free ground to avoid heaving or settling and the cracks that develop from this process. Half-chimneys or chimneys set on wood foundations ideally should not be used.

2. A masonry chimney should have two-inch clearance from wood or other combustibles. Even at floor and roof pierces, this clearance should exist.

3. The mortar should be solid everywhere. An ice pick is a perfect tool for testing mortar. Try to poke it into and through the mortar. If you can, the chimney should be rebuilt to make the mortar solid.

4. The chimney wall should be at least four inches thick with a ⅝-inch liner.

5. The chimney should be capped with a well-sloped and waterproof cap.

6. The height of the chimney should be at least two feet above the roof peak, or above any portion of the building within ten feet.

7. Chimney liners should extend at least eight inches below the connector inlet, preferably all the way to the soot cleanout door. Liners should be separate from the chimney wall. Space between the liner and masonry can be empty or filled with vermiculite.

8. Joints between liner tiles should be built up on the *outside* of the liner. (See illustration on page 10.)

9. Any wood beams resting on or supported by the masonry chimney should have at least eight inches of solid masonry between the wood end and the flue liner.

10. No combustible lathing or furring strips should be placed against a chimney. (The exception to this is that a chimney built entirely on the outside of a building may touch the sheathing.)

Many of these things are impossible to see once the chimney has been built. Others are easy to see. In any event, keep in mind that the inspection should be aimed to provide a safety margin even in the event of a chimney fire.

Cleaning Fireplaces

Cleaning the fireplace is usually the most difficult part of the cleaning job. Most fireplaces are made of brick, but there are also stone and block fireplaces as well as prefabricated fireplaces. Prefabricated electric fireplaces are also available, but these fireplaces do not need cleaning.

Fireplaces are cleaned to remove creosote, just as chimneys are, but creosote accumulates more slowly in the fireplace than it does in the chimney flue. The illustration shows the structure of a typical fireplace. The fireplace confines the fire, and its temperatures are high enough to keep large amounts of creosote from accumulating on its walls. In addition, much of the tar-laden smoke never touches the walls of the fireplace, further reducing the rate of creosote deposit. You may question, then, Why bother to clean a fireplace at all?

THE FIREBOX

The firebox of a circulating air fireplace, such as a *Heatilator* brand, should be kept clean to ensure good heat transfer across the steel wall. Also, a clean fireplace looks better. But the best reason to clean the firebox is to keep yourself a bit cleaner when you do the smoke chamber.

Clean From the Inside

Fireplaces by nature must be cleaned from inside the house. The proximity of rugs and furniture to the very large opening makes a fireplace

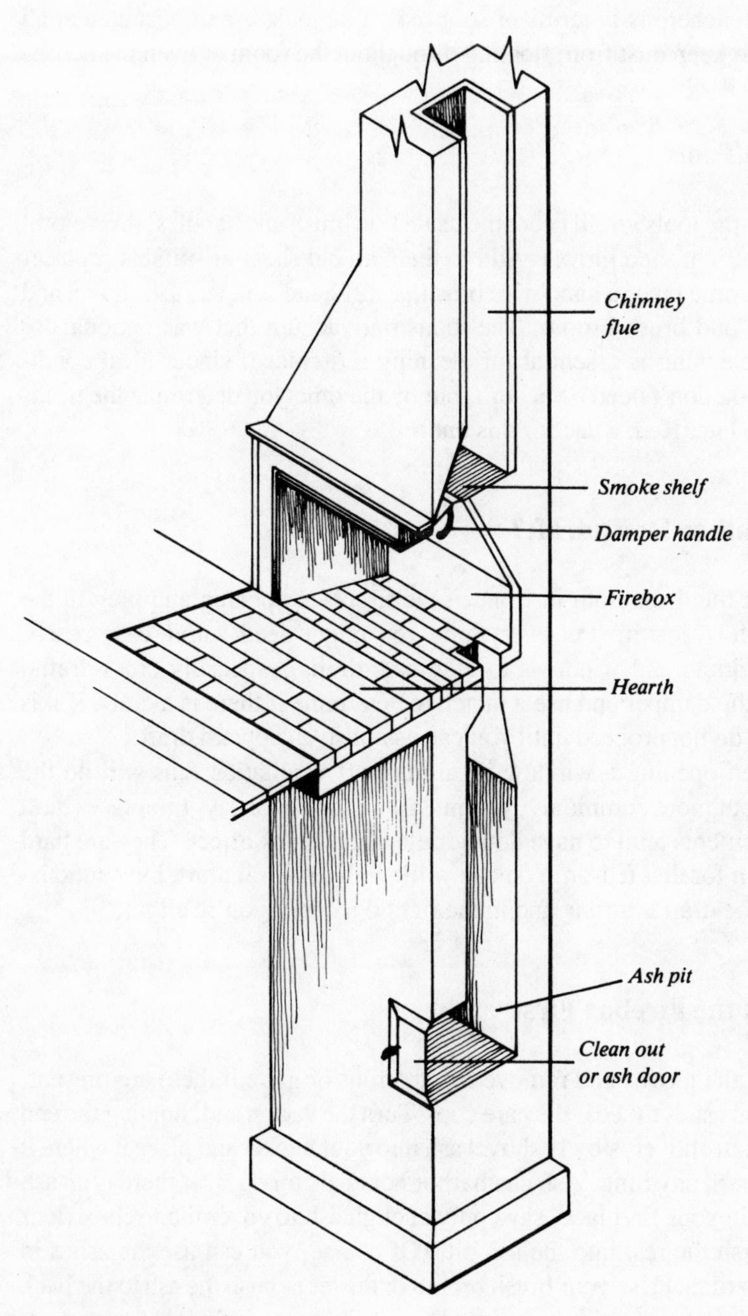

How a fireplace is constructed.

most treacherous in terms of soot risk. You must use utmost care at all times to keep dust from floating throughout the room or even the house.

Extra Tools

Gather the tools you'll need: the usual flue brush and handles, drop cloth, trouble light, and gloves. Add to them an old sheet or suitable replacement, some tape, a hand wire brush, a stiff hand scraper, ash shovel and bucket and brush broom. The industrial vacuum that was optional for stove cleaning is essential for cleaning a fireplace. Under ideal conditions you don't need a vacuum, but by the time you determine the need, it's too late. Rent a vac for this one.

Updraft or Downdraft?

Spread the drop cloth in front of the fireplace opening and plug in the light and vacuum. Put everything you might need within easy reach. Close doors and windows to eliminate drafts, especially cross-drafts. Open the damper and use a match to determine if there is a draft. If it is down, do not proceed until you can establish an upward draft.

Often opening a window or turning off ventilation fans will do the trick, but more commonly, you must wait for a cool day. Fireplaces built in basements tend to have downdrafts due to stack effect. They are hard to clean for that reason. You can work with a neutral draft, but cautiously. If the draft is strong and in the right direction, you're all set.

Clean the Firebox First

Put on the gloves and remove the andirons or grate if these are present. Feel the ashes to see if they are cool. Turn the vac on and, holding the end near your shovel, slowly shovel ash into your bucket and place it where it can't burn anything. Ash can harbor hot coals for days. If there is an ash dump in your fireplace, save only enough ash to cover the firebox floor and push the rest into the ash pit. (Of course, you can use the ashes in your garden.) Use your brush broom or duster to push the ash to the back of the fireplace to aid shoveling. Do not sweep it, as this will raise dust. Drag the brush slowly.

Start a fireplace by shoveling out the ashes. The dropcloth and industrial vacuum will help you keep soot risk to a minimum.

Brushing the fire box. The trouble light will help you do a thorough job.

When this has been done, use the hand wire brush to brush the firebox clean. The corners take extra time, so don't hurry them. Close the damper and brush off its surface, spending extra effort to clean the area where the damper handle attaches to it. The vacuum should be running the whole time, keeping the end of the hose near the brush or shovel. If you're cleaning when the weather is 0° C. or cooler, chances are the draft will be strong enough when the damper is open to eliminate the need for the vac at all. Basement fireplaces are sometimes an exception to this rule.

SMOKE BOX AND FLUE

When the entire firebox is clean—and this does not mean that the original color has been restored, just that no discernable thickness remains—you must decide what approach you'll take for the smoke rise box and flue. Your basic choices are cleaning the flue from the roof or from inside. Take into consideration such things as height of the roof and availability of an appropriate array of ladders, presence or absence of a chimney cap, direction and strength of draft, and size of the fireplace and throat. Cleaning from inside the house generally is easier. If you elect the inside approach, proceed by first removing the damper.

Remove the Damper

Removing the damper is sometimes the most difficult part of the job. The variety of dampers is astonishing. Some fireplaces are fitted with dampers that are not designed to be removed. If you're cleaning only your own chimney, you'll only have to learn how one damper comes out. If fifteen or twenty minutes of effort is not enough to remove your damper, you can assume that it probably will not come out. If it won't, clean the smokebox and flue with it in place.

The first step in removing any damper is to detach the *damper handle*. Damper handles come in a wide variety of designs, most of which are only slightly more complicated than a ball point pen. The trouble is that they are all in awkward positions and usually are covered with creosote and soot. This is why it is important to brush the mechanism with a wire brush *before* attempting to remove it.

A damper, with handle attached. Both should be removed before you clean the chimney flue.

The simplest type of handle is a curved bar, pivoting at one end on the damper plate and having a number of notches or steps that can be rested on a support on the front wall. The support is usually a loop through which the damper handle passes (see illustration.) The handle is fixed to the damper with a cotter pin. Remove the pin and the handle should slide down through its support loop.

Another popular type of damper handle is also shown. Its operation is not quite so simple. Remove the cotter pins at each end only. There is sometimes a C-shaped wire through the cotter pins; note its position to allow easy reassembly.

Notice that the shortest position (which would make the damper closed) leaves the ring handle at the back of the firebox. This is true for most damper handles. They are supposed to be out of sight (e.g., at the rear end of the firebox) when the fireplace is not in operation and the

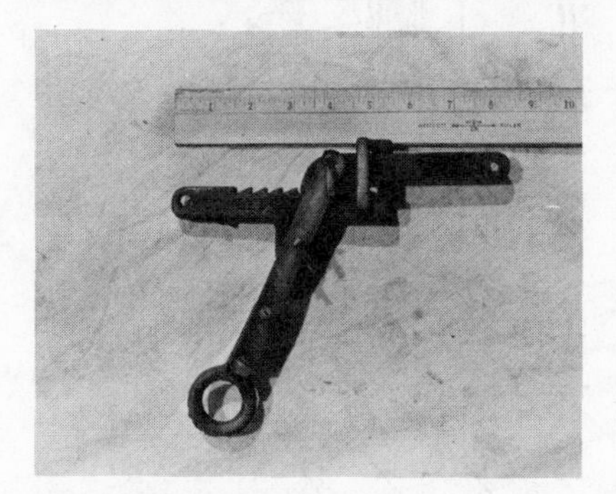

Damper handle in open position, left, and closed, right.

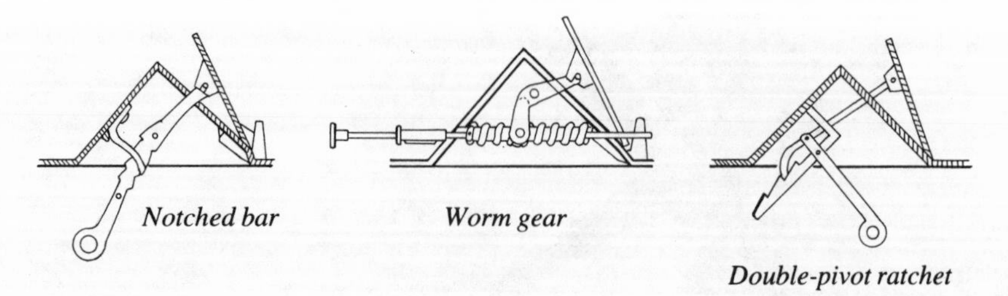

Fireplace dampers.

damper is closed. When you have removed the damper handle mechanism, wire brush it clean.

Some fireplaces have *twist-type controls* on the front of the fireplace, usually just above the opening to the fireplace (see illustration). Typically, a shaft leads through the masonry to a worm gear that, when twisted, drives another piece to lift the damper open. The worm gear usually has a set screw at one end. Use your screwdriver to loosen this screw. When it is loose, you should be able to withdraw the shaft by pulling straight out

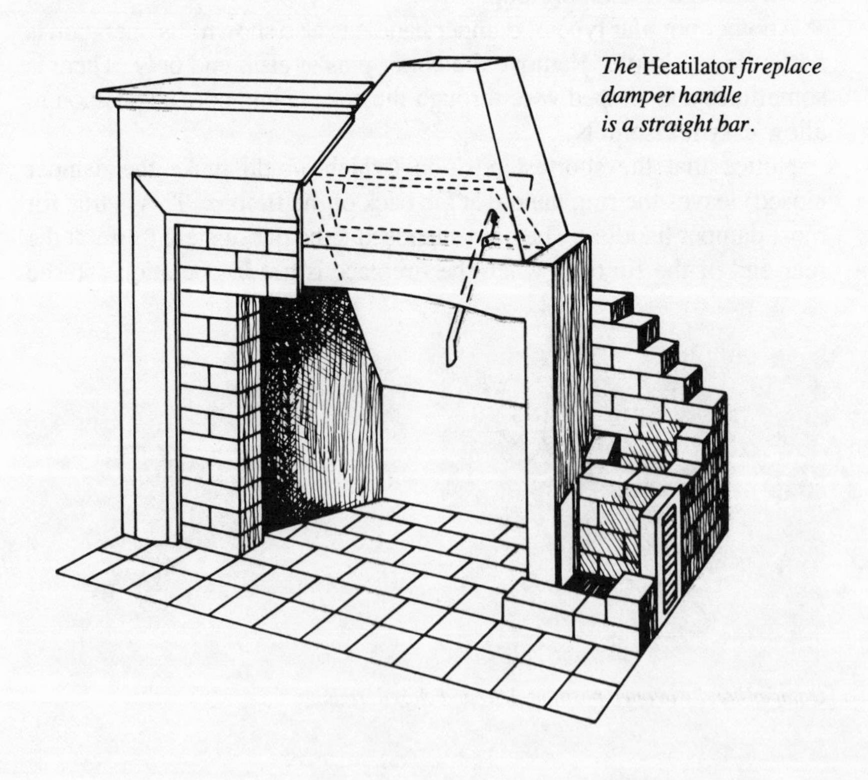

The Heatilator *fireplace damper handle is a straight bar.*

on the handle. The worm gear and the lifting piece will be left hanging from the damper by its cotter pin. Remove this pin and brush the mechanism.

Another common type of handle is a simple bar, pivoted on the rear wall of the firebox. The bar leads to a slotted anchor on the damper. This type of handle is used on many *Heatilator* brand fireplaces. Simply remove the cotter pin and pivot pin, leaving the actual handle bolted to the back wall. There are many other types of handles, most of which reveal their secrets of removal upon examination.

When the damper handle has been removed, remove the damper itself. Dampers are sometimes held in place by their weight only, but more commonly they are held by pins that extend from the lower edge of the damper into fittings on either side of the damper openings. Grasp the damper and try simply lifting it out. If it comes out, wire brush both sides and set it out of the way. If it will not budge, assume it has pins on the ends and try moving it from side to side. Years of accumulation of creosote may clog the motion, so keep at it a while. If diligent effort will not remove the damper, assume it will not come out. Brush it in place, then proceed by cleaning the flue.

Carefully remove the damper and brush it in the firebox.

Clean the Fireplace Flue

First, attach your fluebrush (the proper size and shape, of course) to a handle and force it through the damper opening. Leave it there. Tape an old sheet up in front of the fireplace, sealing it against dust. Do not tape it tightly across the opening, but leave plenty of slack in the fabric. Be sure to tape the bottom edge as well.

Now cut a slit in the sheet large enough to allow your handles to go through, but small enough not to emit too much dust. Attach another handle to the first one and push the brush up into the flue (see photos). Sweep the flue slowly to avoid inducing any downdrafts caused by falling creosote. Keep an eye out for escaping dust. If dust does begin escaping, place the end of the vac hose under the sheet inside the firebox.

Sweep the entire length of the chimney flue and withdraw the brush and handles, leaving the brush and one handle in the smoke chamber. Now the smoke chamber must be cleaned and the smoke shelf emptied.

With damper removed, attach one handle to brush and lodge brush in the flue. Note that the square masonry flue requires a square, rather than round, brush.

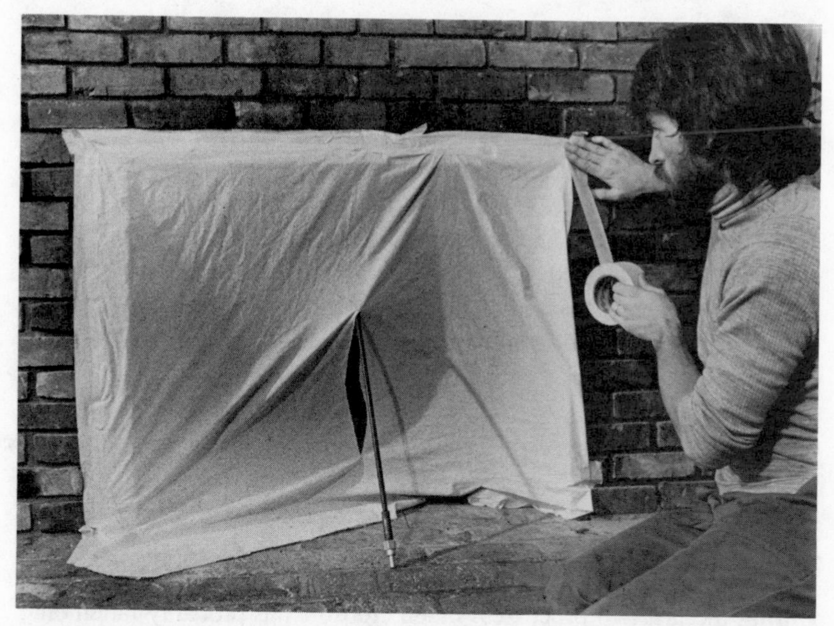

Leaving the brush braced, tape an old sheet over the entire firebox opening, cutting a slit for the handle. Be sure to tape the bottom, too.

Brush the flue, adding handles as you go up.

Clean Smoke Chamber

If you left enough slack in the sheet when taping it, you may be able to leave this shield in place. If not, this job must be done most carefully. Be sure the vac is running and the hose is in the fireplace—or better, in the smoke chamber. Use the flue brush and handle to brush the inside of the smoke chamber as best you can. Remove the brush and handle, and using the drop light, peer up into the chamber. Seek out the spots that need further work and scrub them with the hand wire brush. Smoke chambers are often *corbelled* or stepped in, leaving many spots for creosote to accumulate. Brush each spot clean.

Shovel Off Smoke Shelf

During the last two processes, much creosote has probably fallen onto the smoke shelf. Remove this debris now. Using your small ash shovel, reach up through the throat and shovel all the creosote out. This is always

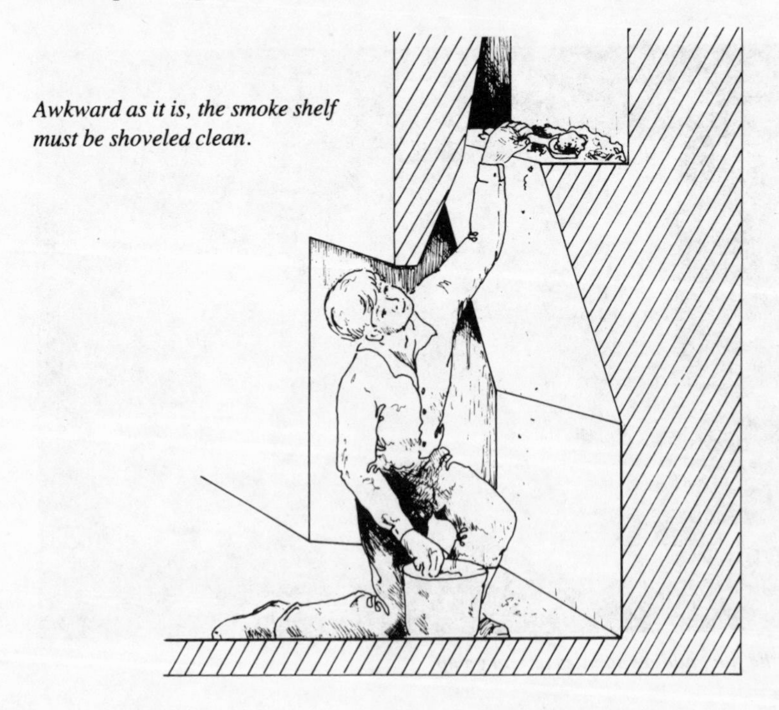

Awkward as it is, the smoke shelf must be shoveled clean.

When the smoke shelf is shoveled as clean as you can get it, use the vac to remove any remaining creosote.

difficult, especially with long-handled ash shovels, but the alternative—sweeping the shelf debris off and onto the firebox floor—generates so much dust that it is nearly impossible to contain. If the damper was left in place, shoveling the smoke shelf clean will prove exceptionally difficult. Nonetheless, the smoke shelf must be cleaned.

When the shovel no longer is of great use, use the vacuum to suck the last bits of creosote from the smoke shelf. This completes the cleaning. All that remains is to replace the damper and handle and place a bed of ashes back in the firebox. These ashes will protect the fireplace floor and aid in keeping fires going.

Cleaning from Outside

If you elected to clean from the outside, you save the hassle of taping the opening (that is, of course, if your firebox has a reasonably tight damper). First, close the damper. Then go up the chimney top and sweep down from the top. Again, sweep slowly to avoid the downdraft.

Sweeping a masonry chimney from the roof, demonstrated here by author Chris Curtis.

When the flue is done, return inside and remove the damper handle and damper. Clean the smoke rise box and smoke shelf. Replace the damper and handle. Replace the ash, and clean up.

Special Cases

Not all fireplaces have dampers or even smoke shelves. If they do not, just clean whatever you find. Fire screens and glass doors are a problem to work around, but they just require a bit more patience.

Prefabricated Fireplaces

Prefabricated fireplaces are generally easier to clean. They typically have smooth metal interior walls that are easy to brush. Dampers are generally round plate types in the chimneys, which are usually round metal. There commonly is no smoke shelf to worry about.

To clean this type of fireplace, first clean the firebox as outlined before. If there is a plate damper, it usually is easier to brush the chimney flue from the roof. Again, close the damper before doing this. When the flue has been brushed, return inside. Start up the vacuum and, with the hose end in hand, slowly open the damper. Put the hose near where the creosote and soot lands, so that the dust is sucked into the vac. When this is done, replace some of the ash on the firebox floor and pick up.

Check the Fireplace for Safety

When your fireplace is clean, it is wise to inspect it for safety. See the safety guidelines.

GUIDELINES FOR FIREPLACE SAFETY

1. There should be no cracks in any of the masonry, especially in the firebox.

2. The hearth should be in good condition and should not rest on or be supported by wood or other combustibles. It should extend at least twelve inches on either side of the firebox and eighteen inches in front of it.

3. The ash cleanout door should fit tightly.

4. Fireboxes should be firebrick lined.

5. A mantle should be at least six inches from the fireplace opening. If it projects more than one and a half inches from the face of the fireplace, it should have at least a twelve-inch clearance.

6. Fireplace openings should have fire screens.

7. Fireplace chimneys should have two inches of clearance on all sides. Fireplace backs should have a four-inch space between the masonry and combustibles.

8. Prefabricated fireplaces often depend on thin metal walls and air circulation to keep nearby combustibles safe. Corrosion is an indication of unsafe spots. Ash and creosote must not restrict free air flow around these appliances.

If you have followed the directions in this book carefully, you should have a clean chimney as well as a clean house. You may be a bit dirty, but you should also be better acquainted with your woodburning system and how to keep it safe. Regardless of the outcome of your endeavors, you've now been introduced to the world of chimney sweeping. You may wish never to try it again, or perhaps you've taken pleasure in the art, even caught a bit of the magic and will look forward to the time you'll do it again.

Hiring a Professional Sweep

If you've read this book and decided to leave the task to professionals, or tried it once and found it not your cup of tea, here are some things to bear in mind when you hire a professional sweep:

1. Call a few to get estimates. If you can get a firm price on the phone, so much the better.

2. When you speak with him or her, evaluate his phone manner. Someone who is courteous and thorough on the phone is apt to be the same in person.

3. Find his or her level of knowledge. Ask specifically what he does. After reading this book you should be able to make up a few questions that will give you an idea about his knowledge.

4. Ask about the mess. Does he guarantee no mess?

5. See if he carries insurance; any good sweep carries a lot.

6. Ask if he has had any formal training in chimney sweeping. If so, where did he train? Did he go to school or merely watch another sweep for a while?

7. What kind of equipment does he use? Black Magic has a great set of equipment; so does August West. Beware of the homemade system (although some of these work very well).

8. Don't always pick the cheapest bidder. You get what you pay for.

The professional sweep in his element.

Glossary

Ash Dump: A hole in the floor of a fireplace that can be used to dump ashes into the base of the chimney.

Ash Dump Cover: A cover for the ash dump.

Ash Pit: The chamber in the base of a chimney into which ashes are dumped, affectionately called an *ash hole*.

Btu: British Thermal Unit—the amount of heat needed to raise 1 pound of water 1° F.

Chimney: The part of the smoke handling system that is built into the structure it serves. It may be brick, concrete, metal or ceramic.

Chimney Connector: The part of a smoke handling system that carries the smoke from a stove to a chimney; a stovepipe.

Cleanout Door: The door at the base of a chimney used to clean out soot, ash and creosote.

Corbel: A mansonry term meaning to step in each course of bricks.

Damper: A device inserted in a chimney or chimney connector or stove to reduce the draft on the fire.

Downdraft Equalizer: A device inserted into a stovepipe to reduce downdrafts.

Draft Controls: Control, usually on the upstream side of the fire, used to limit the air available to the fire.

Firebox: The chamber of a stove or fireplace where the fire burns.

Fireplace: An open burning appliance, sometimes made of brick, and built into a brick chimney.

Flue: The interior chamber of the chimney extending from the smoke chamber to the sky.

Hearth: The floor of a fireplace that extends into the room.

Heat Reclaimer: A stovepipe device to recover heat that would normally be lost.

Smoke Chamber: The chamber in a fireplace immediately above the throat and smoke shelf. Also *smoke dome, smoke rise box.*

Smoke Pipe: Chimney connector; stovepipe.

Smoke Shelf: A more or less flat shelf in a fireplace behind the damper. It forms the floor of the smoke chamber.

Stack: A chimney, sometimes the whole smoke handling system.

Stovepipe: Thinwall pipe used to connect a stove to a chimney. Also *smoke-pipe* or *chimney connector*.

Thimble: A device to allow safe passage of a stovepipe through a wall or ceiling.

Throat: The narrow passage in a fireplace between the firebox and smoke chamber.

Tiles: Flue tiles, a ceramic tile liner for chimneys.

Index

Other Garden Way Books You Will Enjoy

If energy conservation interests you, we're sure you will like the following other Garden Way books:

Wood Energy: A Practical Guide to Heating With Wood, by Mary Twitchell. Quality paperback, 8½ × 11, 176 pages, $7.95. The definitive wood heat book, with comprehensive catalog section on stoves and furnaces.

Home Energy for the Eighties, by Ralph Wolfe and Peter Clegg. Quality paperback, 8½ × 11, 288 pages, $10.95. How to deal with the energy crisis by turning to solar heat, water power, wind power, and wood. Plus incredible catalog sections on what's available now in these fields.

Wood Heat Safety, by Jay Shelton. Quality paperback, 208 pp., 8½ × 11, $8.95. Complete illustrated reference to wood stove and chimney safety, covering all aspects of safe installation and operation. Essential for chimney sweeps.

Build Your Own Low-Cost Log Home, by Roger E. Hard. 204 pages, 8½ × 11, quality paperback, $6.95; cloth, $10.95. A remarkable complete home construction book.

Designing and Building a Solar House, by Donald Watson. 288 pages, 8½ × 11, quality paperback, $8.95; cloth, $12.95. 'A nuts-and-bolts book that brings the sun down to earth,'' said Alvin Toffler, author of **Future Shock.**

At Home in the Sun: An Open-House Tour of Solar Homes in the United States, by Norah Davis and Linda Lindsey. Quality paperback, 8½ × 11, 248 pages, $9.95. What it's like to live in a solar house, as told by the owners of thirty-one solar homes around the country. Heavily illustrated, plus technical information.

Harnessing Water Power for Home Energy, by Dermot McGuigan. 112 pages, quality paperback, $4.95; hardback $9.95. An authoritative, detailed look at the uses of small-scale water power.

Harnessing Wind Power for Home Energy, by Dermot McGuigan. 144 pages, quality paperback, $4.95; hardback $9.95. A solid, complete analysis of wind power options for homeowners, with details on machines, manufacturers, and whole systems.

These books are available at your bookstore, or directly from Garden Way Publishing, 171X, Charlotte, Vermont 05445. If ordering by mail and your order is under $10, please enclose 75¢ for postage and handling.